Published by
Saurabh Publishing House

Psychology of Learning and Human Development

M.L. Ojha

Saurabh
Publishing House
4735/22, Prakash Deep Building,
Ansari Road, Daryaganj,
New Delhi-110002

Published by :
SAURABH PUBLISHING HOUSE

Distributed by :
LOTUS PRESS PUBLISHERS & DISTRIBUTORS
Unit No. 220, Second Floor, 4735/22, Prakash Deep Building,
Ansari Road, Daryaganj, New Delhi - 110002
Ph : 011-23280047, 32903912, 098118-38000
E-mail : lotus_press@sify.com

Saurabh Publishing is an imprint of
Lotus Press Publishers & Distributors

Psychology of Learning and Human Development

ISBN : 978-93-83045-04-4 (H/B)

Printed at : Bharat Offset Works, Delhi

Preface

The school as an institution exists to guide children in the development of skills, subject-matter, attitudes and behaviour patterns essential for effective living in a particular society. The teacher, therefore, has two major tasks: to help children learn as efficiently as possible the subject-matter being studied and to help them acquire those desirable personal and social behaviour patterns. Where religious education is part of the curriculum, the teacher's function is to nurture the child in the proper relationship to God. Since children need the teacher's guidance to learn and since teaching may be defined as a set of actions directed to pupils, an important aim of educational psychology is to help teachers acquire understandings needed to guide children in their educational development. Teaching is based upon the concept that children can learn. Although many studies dealing with the learning ability of children at different maturational stages have been conducted, considerable theoretical differences exist about the relative importance of heredity and environment in learning. Despite numerous studies this problem is still unresolved. The viewpoint presented throughout this book, however, is that the child is a product of the interaction of the human organism and various environmental forces and conditions. This viewpoint emphasizes the role of both heredity and environment—not one to the exclusion of the other. Certain principles have emerged from this viewpoint, an understanding of which is essential for a successful educational program.

The science of psychology has not yet reached the stage of maturity whereby it can present a unified, coherent and generally acceptable set of principles of learning, motivation and adjustment. However, scientific studies have furnished useful information to teachers and others concerned with guiding and motivating pupils in learning situations. Educational psychology has furnished teachers and students of education with the informational background and scientific methods for studying pupils and evaluating different educational practices. This book incorporates recent research on problems dealing with children and adolescents in learning situations. Such research will provide us with additional information and encourage keener insight into many educational problems.

This book deals with the topics of—*Nature of Educational Psychology; Human Growth and Development; Cognitive Development; Social and Emotional Development; Learning—Laws and Theory; Intelligence and Creativity; Motivation and Group Dynamics; Personality and Character Assessment; Delinquency, Exceptional and Backward Children; Guidance and Counselling;* etc.

I hope that students and those in the education field will find it useful for academic and competitive examinations. I owe a deep sense of gratitude to my publisher for materializing my present endeavour.

M.L. Ojha

Contents

1

Nature of Educational Psychology

The word Psychology is derived from two Greek words, 'Psyche' and 'Logos', "Psyche' means soul and 'Logos' means science. Hence, the original or etymological meaning of Psychology is the science of soul. In due course the meanings of psychology have undergone many changes. From Science of Soul, it became 'Science of Mind', 'Science of Consciousness' and 'Science of Behaviour' which is its modern concept.

1. Psychology as a Science of Mind. Philosophers in the middle ages considered psychology as the science of mind. They used to distinguish between the mental and spiritual life of the individual. These philosophers, however, failed to give the exact nature and, force of the mind of the individual. Hence, psychology the science of mind could not progress.

2. Psychology as a Science of Behaviour. The latest and modern concept of Psychology is in times of behaviour. Behaviour is the resultant manifestation of both conscious and unconscious urges. Behaviour can be observed objectively and is a more suitable method of studying Psychology.

3. Psychology as a Science of Soul. The earliest definition of psychology was 'a Science of Soul'. Philosophers like Plato, Aristotle and Descartes interpreted psychology, according to this concept. In the 16th century this meaning of psychology was rejected.

4. Psychology as a Science of Consciousness. In the 19th century psychologists like William James, William Wondt and others considered psychology as a science of consciousness. By consciousness the psychologists meant awareness or wakefulness. They, however, had different interpretations of the term wakefulness. As this definition of psychology had a very limited approach, it was not universally accepted.

Modern Definitions of Psychology. Let us now acquaint ourselves with some of the definitions of Psychology:

1. 'Psychology is the study of human nature'. (**Boring Lang Feld**).
2. 'Psychology is the study of human behaviour and human relationships'. (**Crow and Crow**).
3. 'Psychology is the science of behaviour and experience'. (**Skinner**).
4. Psychology undertakes to make a scientific study of the individual considered as a unit as he really is in his dealings with other individuals and with the world.' (**Woodworth**).

The most acceptable Meaning and Definition of Psychology. It is in terms of behaviour. A lot of work has been done by psychologists in this field. Detailed experiments have been conducted on human behaviour. William Wondt established the first Psychological Laboratory at Leipzig in 1879.

Human behaviour is highly complex. All human beings do not behave in the same manner. They differ in abilities, likes-dislikes, interests, attitudes, emotions and sentiments. Why is it so? The science that answers such questions and studies human behaviour is called Psychology:

Classification of Psychology. Psychology can be divided into eleven branches:

1. General Psychology. It studies the behaviour of human organism in general. Its scope is very large but generally it studies normal organisms. It is the mother of all other branches of Psychology.

2. Adult Psychology. It studies the behaviour of adult humans. An adult is a mature person and his emotional, social and intellectual behaviour is not like the behaviour of a child. So it is totally a separate branch of Psychology.

3. Abnormal Psychology. The scope of Abnormal Psychology is limited to the study of abnormal individual only.

4. Individual Psychology. It deals with the variation in human beings. No two persons are alike. They always differ in their behaviour according to their intelligence, race and sex. They also differ in behaviour in other factors such as interest, philosophy and education.

5. Child Psychology. It deals with the development of behaviour of the child. The child is an adult in the making. He differs from the adults very much. "His life urges, his emotions, his sentiments, his intelligence and his aspirations all differ from those of the adult individuals. Therefore, Child Psychology is a fullfledged science in itself.

6. Social or Group Psychology. This branch of Psychology deals with the behaviour of individual as member of a group or mob does not act in the same way as a majority of the individual members would act individually.

7. Industrial Psychology. This is also a branch of applied Psychology. It is in fact Social Psychology with reference to work, individual as well as collective. Industrial work may be made more attractive and interesting and output of industry may be increased and improved if we exploit the finding of Industrial Psychology.

8. Educational Psychology. It is the most important branch of applied Psychology. It is the study of the psychological aspects of educational situations. It is the study of educational problems with reference to psychological facts. Psychology is science of behaviour and Education aims at modifying the behaviour in the most desirable way. But modification of behaviour depends on some fundamental psychological laws and limitations. Educational psychology studies those facts and limitations. It covers the development of the child from early childhood to maturity, general facts of psychology which may have any relation with the modification of the behaviour of the child, psychologised methods of learning, measurement of capacities, attitudes and interests and other problems of applied Psychology related to education.

9. Experimental Psychology. It studies mental processes and behaviour in laboratories with the help of experiments.

10. Para-Psychology. This is the latest development in the field of psychology. It deals with the problems of what happens to an individual after death.

NATURE OF EDUCATIONAL PSYCHOLOGY

Nature of educational psychology is scientific, as it uses scientific methods. Following are the essential elements of science:

1. **Scientific Method.** A science makes use of the scientific method.
2. **Factuality.** Science is the study of facts, the search for true facts.
3. **Universality.** Scientific laws are universal.
4. **Discovery of Cause and Effect Relationship.** Science studies cause and effect relationship, it searches for cause and effect relationship in its material. On the basis of such relationships it creates universal and valid laws.
5. **Prediction.** Science makes predictions.
6. **Validity.** Scientific laws are also valid. They are valid at all times and in all places. They are found to be invariably true. They are open to examination at all times.

On the basis of above criteria of science we find that educational psychology is a science as it possesses all the characteristics of being a science.

The nature of educational psychology is scientific as it employs scientific method in the study of behaviour of the learner in an educational environment. Educational psychology makes use of scientific techniques such as observation, experimentation, case history, interview and projective techniques. Thus, educational psychology as a science is systematic as its procedures are orderly.

Educational Psychology is Empirical and Factual

- It is a body of facts or information that has resulted from observations and investigations. Its conclusions are objective and unperjudiced.
- It has developed a methodology by which investigations are made, information is discovered hypothesis tested and theories are derived.
- It is one of the applied branches of psychology. It studies the behaviour of the learner in relation to learning situation.
- After establishing cause and effect relationships concerning the behaviour of learners it becomes possible for the educational psychologist to predict the behaviour of other learners. He can predict the development of learners in future.
- It is a positive science of human behaviour. Its function is to describe, control and predict human behaviour. Human behaviour is quite dynamic and unpredictable.
- It defines cause and effect relationship and on the basis of this knowledge it analyses behaviour of the learner.

It is clear that educational psychology is an applied and positive science of behaviour of the learner. Its subject-matter is subtle, complex and varied. As the human behaviour is unpredictable, more variable and less reliable, viewing this concept, we can say that educational psychology is not a perfect science. Educational Psychology cannot claim the same degree of validity, accuracy, objectivity and exactness as claimed by natural science.

Aims of Educational Psychology

Aims of educational psychology are parallel with aims of education. Various views have been expressed by different thinkers about these aims as given below:

According to Garrison, "The aim of educational psychology is to know the behaviour of the child and to predict and control it."

According to Kuppuswamy, "The aim of educational psychology is to use psychological principles for good teaching and learning."

According to W.A. Kelly, the following are the aims often subject to:

(1) To give an insight with teacher about the nature of child.

(2) Giving knowledge about the character formation of the child.

(3) Helping the teacher in finding out ways and means of social adjustment for the child.

(4) Enlightening the teacher with the growth and development of the child.

(5) Imparting knowledge about the principle and methods of learning.

(6) A study of the techniques of evaluation of various school subjects.

(7) A study of the psychological methods of investigation used in the field of educational psychology.

(8) Studying about the control of emotions and their educational importance.

Charles Skinner has classified the aims of educational psychology as general and particular:

The gerneral aim is to provide a body of organised facts and generalisations that will enable the teacher to realise increasingly both cultural and professional objective:

The particular aims are as follows:

(1) Creating confidence in the teacher to plan harmonious development of the child.

(2) Understanding the social problems of children and planning for their total social adjustment.

(3) Approaching the problems of children in a sympathetic and impartial manner.

(4) Imparting knowledge to the teacher about the latest developments in the field of child psychology and various methods of teaching.

(5) Enabling the teacher to evaluate himself and his methods of teaching.

(6) Helping the teacher to understand and analyse the behaviour of children.

(7) Planning education of children according to their age, ability and aptitude.

(8) Helping the teacher in solving the problems that may arise in the classroom.

In other words we can say that the general aim of educational psychology is to provide a body of facts and principles which will enable the teacher to enhance and enrich his own life and which will help him to bring about improvement in the quality of instruction, thus, equipping him adequately, both culturally and professionally and the specific aims are following:

(*a*) To explain the nature and characteristics of the learner.

(*b*) To describe the nature of learning processes.

(*c*) To explain the manner in which these processes may be facilitated by the teacher and in the light of these, to explain, how he or she should teach.

(*d*) To assist the teacher in the better adjustment of children and in the prevention of maladjustments.

(*e*) "To assist in defining and setting-up educational objectives and standards in terms of desirable behaviour (conduct, attitudes etc.,) that ought to be the goals of all teaching efforts. If the teacher knows what the desired outcomes should be, he can set-up appropriate situations (curricular materials, teaching procedures etc.) for bringing about the wanted changes." (Skinner)

(*f*) To help the teacher in defining and planning suitable guidance programmes and functional forms of organization and administration.

(*g*) To help in the evaluation of the outcome of the educational process.

(*h*) To aid in developing a scientific problem solving attitude toward the problems of learning and teaching.

Relation Between Psychology and Education

(1) Education determines aims and ideals and psychology decides which aims can be attained.

(2) Psychology deals with facts as they are. Education determines the standards to be achieved. Psychology tells the way of attaining these standards.

(3) Psychology helps the teacher to know the child and plan for the maximum development of the child, which is the ultimate goal of education.

Importance of Educational Psychology to the Teachers

1. Psychology Helps the Teacher in the Realisation of the Aims of Education. The main aim of education is the harmonious and all-round development of the educand. Development comes from within and the inner potentialities must be understood and approached for the desired modification in the development of the child. The knowledge of Psychology helps the teacher to understand the inner potentialities which underlie the behaviour of the child. He can direct and modify the development of the educand quite efficiently if he knows Psychology.

2. Educational Psychology Proves Useful in the Methods of Teaching. Psychology has discovered many secrets about mental forces. Recent researches made in the field of psychology gives us valuable suggestions for the better methods of teaching and memorizing, for making desirable habits and eradicating anti-social tendencies. Psychology tells us how significant play and recreation are for the children and how play-way-methods turn the work of learning into an interesting play.

3. Character Development. Educational psychology helps a lot to the formation and development of character. The teacher comes to know the methods he should adopt in inculcating character traits and moral principles among the children.

4. The Knowledge about the Child is as Necessary as that of the Subject. The educator will have to consider the mental processes of the child and not only the quality of knowledge which he is going to be put into the living mind. Psychology helps him in understanding 'John'. The teacher may be a great scholar but he will not succeed in teaching if he has no knowledge of the mental forces of the child.

5. Psychological Measurements are Very Useful for the Teacher. Psychology has developed many reliable tests and instruments of mental measurement. These are providing to be extremely useful in the field of education. We can quite easily measure mental capacities, basic intelligence, temperamental attitudes and special inclinations of children and base their educational programmes on these findings. These measurements show that all the children differ in their eventual capacities and that every child is a unique being. The teacher can know that children with I.Q. below 90 cannot do well in medical, engineering, administrative or other similar vocations. But he knows that such young ones are not doomed if they cannot do well in intellectual callings. He can easily explore some other fields where such children can also flourish.

6. Educational Psychology and Nervous System. The entire education depends on the function of the brain and nervous system. It becomes essential to teacher to study the nervous system which controls human behaviour. He must have the knowledge of sensory organs which are the gates of knowledge.

7. Knowledge of Educational Psychology will Give a Scientific Attitude to the Teachers. The knowledge of psychology will make an educator more scientific in his educational practices and consequently he will be more and more objective and methodical in his work.

8. Educational Psychology Helps the Teacher in His own Mental Adjustment. A maladjusted personality can never be a good model for hundreds of developing personalities. He cannot solve his own problems and therefore he cannot help the children in this aspect. The study of psychology may help the teacher to readjust himself. He will know his own weaknesses and merits and will overcome his complexes.

9. Educational Psychology can Change the Outlook and Attitude of the Teacher towards his Pupils. The knowledge of psychology will change the outlook and attitude of the teacher towards children. He will no more denounce his pupils on account of their undesirable behaviour. He will not hate the juvenile delinquent. Instead of being fed-up with the backward child the teacher will be more interested in him. The problem child will no more be a problem for him but he will be simply a child with problems and the teacher will be ever ready to help him in the solution of his problems.

1. Educational Psychology and Process of Learning. Teaching and learning go side by side. All education depends upon the learning of new responses and the capacity of a human child to learn new responses. Educational Psychology discusses the nature of learning theories and types of learning for different age levels and situations. It tells that learning is not possible if the child does not co-operate in the learning process. Therefore, knowledge of Psychology becomes essential to a teacher to study all these problems.

11. Educational Psychology and Play. Play is a natural tendency having great educational quality. Psychology helps in playgrounds also. Child finds outlet for pent-up feelings and energy through play. It also gives joy and pleasure to the pupils. Hence, there is a great importance of play-way in education.

12. Individual Differences. No two persons are exactly alike. There is individual difference. Pupils always differ in their level of intelligence, aptitudes, likes and dislikes and in other propensities and potentialities. Different minds are to be trained by the teacher. There are gifted, backward, retarded, talented and handicapped children. All of them should not be treated in the same manner. So, the teacher should know their Psychology to satisfy them.

To sum-up, Educational Psychology has revolutionised education. It has made education child-centred. It is only with the knowledge of Educational Psychology that it is possible for the teachers to render proper guidance, arrange the learning situations and plan the instructional programmes and other activities for the all-round development of children.

PSYCHOLOGY ITS RELATIONSHIP WITH EDUCATION

Psychology has assumed an extraordinary importance in the life of an individual. Infact, every action of life, may it be home, school, neighbourhood or social gathering, is influenced by it. All communications between the teacher and students are based on the knowledge of psychology. Theories of intelligence, creativity and personality have been responsible for shaping and designing the educational system according to individual needs.

Psychology is the Science of Soul. The word 'psychology' comes from Greek words 'Psyche' and 'logos' where psyche means soul and logos means science. Thus, literally psychology means a science of soul. But this is not acceptable because of following criticisms:

(i) Nature, origin and place of soul is not known. It has no physical existence. It cannot be seen and heard. It has no weight and volume. It is a metaphysical cancept.

(ii) Soul makes science speculative as it is not open to observation and experimentation and therefore, not given to verification. So this definition of psychology is unscientific and hence, this is discarded.

(iii) It makes psychology more of religion than of science because it is a theological concept and implies certain theories of religion and relationship to God.

Psychology is Science of Mind. Alternatively philosophers evolved another definition of psychology as science of mind. Mind is a combination of sum total of mental processes and it stands for personal internal experiences of man, *e.g.*, pleasures and pains, hopes and wishes, dreams and desires. But this is also criticised on following grounds:

(i) It is subjective. We can know our own mind and not the mind of others. So it is half true.

(ii) Mind implies continuity and unity but it is lacking in abnormals, insanes and animals etc.

Psychology as a Science of Consciousness. Failing to define the nature of soul and mind, the psychologists defined psychology as a science of consciousness. The definition was considered too narrow as this did not include the unconscious and pre-conscious activities of mind. Freud in his psycho analytical theory described that mind consists of 3 states—pre-conscious, unconscious and conscious. About $9/10^{th}$ part of behaviour is determined by unconscious and pre-conscious mind whereas only $1/10^{th}$ part is determined by conscious mind. But in Freud's terminology, unconscious mind plays a major role in the development of personality of an individual. This definition was also discarded.

Psychology as a Science of Behaviour. Psychology was finally defined as the science of behaviour where the term behaviour carries a very wide meaning. According to Woodworth, "Any manifestation of life is activity" and behaviour is a collective name of all such activities. It includes all types of activities such as cognitive, affective and conative. So behaviour refers to entire life activities and experiences of all living organisms.

According to Garrison, "Psychology is concerned with observable human behaviour."

According to Crow and Crow, "Psychology is the study of human behaviour and human relationship."

According to Munns, "Psychology today concerns itself with the scientific investigation of behaviour."

According to Skinner, "Psychology deals with responses to any and every kind of situation that life presents by adjustment activities."

Thus, it can be concluded that psychology is a behavioural science which helps us to describe, understand, predict and control behaviour.

PROBLEMS OF EDUCATIONAL PSYCHOLOGY

1. Development of a Balanced Personality. The main problem that engages the attention of educational psychology is the development of a balanced personality in the child. And, generally speaking, a balanced personality is one in whom the intellectual, emotional, ethical and moral aspects are properly blended and synthesised, functioning properly.

2. Motivation. Often enough, the problems that are at the root of various adjustment defects are problems of motivation.

3. Individual Importance of Heredity and Environment in Individual Differences. The aim of education is the development of the child. Hence, one of the important problems facing the educational psychologist is the determination of the relative importance of heredity and environment in the child's development and in causing individual differences among children. Can education change the inherited qualities? Can children reach an adequate stage of development on their inherited qualities alone if they are placed in a common environment? Is it possible to make a child into anything on the basis of education irrespective of his heredity? These are questions of this nature that encourage and inspire the educational psychologist to make a comparative study of hereditary and environmental factors. *For example,* one important question is the extent to which intelligence can be developed through education regarding which educational psychologists have discovered numerous facts after much investigation.

4. Increase in Understanding and Meaning. Another factor at the root of the motivation problem is sometimes a deficient understanding and an inability to comprehend the implications of the problems existing. It is to this end that the material chosen in teaching nowadays is such as is drawn from an environment with which the child is reasonably familiar and with which he is conversant enough not to be embarrassed by it. A simple example would be the creation of a model government within the class in order to explain the working of the actual government of the country to the student. In democratic societies the students usually understand the functioning of a democratic system in their student groups.

5. Measurement and Development of Intelligence and Attainment. Education develops intelligence and attainment. Educational psychologists search for and discover methods and measures whereby one can know how this development can be achieved and once it has been achieved, to know how much development in the child has taken place.

6. Education of Emotions. The proper and adequate development of personality demands the proper and complete development of emotions. An absence of such education of emotions results in many mental diseases such as indiscipline, split personality, etc.

7. Methods for Maximum Progress in Learning. Fundamentally, the process of education advances with learning. The most prominent question that interests the educational psychologist in this regard in what methods of learning can most facilitate learning and achieve the maximum progress? One such question could be—is there any conflict between imitation and creation? Should education pay more attention to the collection of factual details or concentrate on creative activities? Psychology has made considerable experiments with such methods as imitation, conditioned response, trail and error and insight and has tried to discover the conditions that are best attend to the application of these methods, the co-ordination between the condition and method being productive of the best results.

8. Individual Differences. Practical psychology is based on the fact of individual differences. Educational psychology, a knowledge that makes practical use of psychology in the field of education, finds out why and in what respects individual beings differ from each other. These differences are so obtrusive and well defined that the skilled teacher cannot function properly unless he can understand them. *For example,* the mental level of some individuals is so low that they can be taught even fundamental activities such as wearing clothes and eating only with the greatest of difficulty, while some children, precocious ones, are so intelligent and quick-minded that the normal teaching and study level of the class is not enough to occupy them so that they soon tire of it. Hence, the educational psychologist attempts to discover the limits, if any, of these individual differences and even evolves methods for their measurement in exceptional children. The modern teacher discovers these individual differences and then makes efforts to help the children progress individually.

9. Problems Relating to Learning. Many of the problems of educational psychology are concerned with learning. The main among them are:

(a) Nature of the process of learning.

(b) Importance of individual differences to the speed of learning and limits of learning.

(c) Relative importance of the formal and informal methods of the education or learning.

(d) Relation between level of maturity and learning.

(e) Modes of evaluating the progress of learning.

(f) Internal changes taking place within the process of learning.

(g) Relation between the methods of teaching and the results of learning.

(h) Psychological effect of social conditions upon the learner.

10. Practical Problems. Educational psychology is no less a practical science than it is a theroretical one. Hence, in it one comes across many problems that arise before the teacher when he is teaching normally. Such problems would be:

(i) Individual and social adaptation necessary for adjustment in modern culture.

(ii) Knowledge of the nature of every student and of the possibilities of his development.

(iii) Search for the fundamental psychological causes at the root of the processes of the learning and teaching.

As Educational psychology progresses, it comes face to face with newer problems. In modern democratic societies, educational psychology is engaged in adapting education to democracy. And it has the services of psychology on the one hand and many other social sciences on the other at its disposal in the solving of any problems that it come across in such an effort. And it is on the solution of these problems that the future of democratic societies depends.

SCOPE OF EDUCATIONAL PSYCHOLOGY

Scope refers to the fields of operations, applications, topics and subject matter with which it deals. Since educational psychology is an emerging and growing subject, its scope has never been fixed or static. It has always been changing its boundaries. As a result of new findings in the field, educational psychology has always been adding new dimensions to its areas of activity.

According to Douglas and Holland, "The subject-matter of educational psychology is the nature, mental life and the behaviour of the individual undergoing the process of education."

According to Crow and Crow, "Educational psychology is concerned with the conditions that affect learning."

It has been correctly said by Garrison and others that "The subject matter of educational psychology is designed to enhance and enrich the lives of the learners and to furnish teachers with the knowledge and understanding that will help them to bring improvements in the quality of instruction.

The following areas can be included in the scope of educational psychology:

Study of Heredity and Environment. Educational psychology studies the impact of heredity and environment on the student.

Study of Stages of Child Development. Educational psychology studies the nature of growth and development and general characteristics of physical, mental, emotional and social development at the various stages, *i.e.*, at the stage of infancy, late childhood and adolescence.

Study of Development of Various Aspects of Child. Educational psychology enquires how various modifications can be brought about in the behaviour of the child, how and when do physical, social, emotional and language development occur in children and how do various instincts make their appearance.

Study of Motivation. Educational psychology studies the nature, principles and techniques of motivation in educational field.

Study of Psychological Basis of Behaviour. It studies the psychological as well as the physiological behaviour of the learner. Psychological basis of behaviour are instincts, emotions, sentiments, suggestion, sympathy, imitation and play etc. Physiological basis of behaviour are nervous system and glands etc.

Study of Learning Process. Educational psychology also studies nature and process of learning, laws and methods of learning as well as factors of learning. Interest, attention, motivation, memory, habits and transfer of training play very significant role in leraning.

Study of Individual Differences. Educational psychology studies individual differences and suggests ways and means to provide education to all types of pupils. No two individuals are alike. Individuals differ physically and psychologically. Individual differences are mainly caused by heredity and environment.

Study of Intelligence and Personality. Educational psychology also studies nature, development and assessment of personality and intelligence.

Study of Exceptional Children. Educational psychology studies about exceptional children. It studies about backward, retarded, gifted and creative children and their adjustment problems with home, family and school.

Study of Measurement and Evaluation. Educational psychology is helpful in giving newer and newer techniques and tests for measuring intelligence, personality, aptitudes, interests and achievements etc. It is educational statistics which helps us in measurement and evaluation.

Study of Group Dynamics. Educational psychology studies group dynamics and group behaviour in classroom teaching and learning. It studies importance and effect of group on the individual. It also considers how an individual can modify the group life. It emphasises the role of groups in the education of children.

Study of Guidance and Counselling. Child is like a book whose every page has to be studied by the teacher. The teacher has to guide and counsel the students at every stage. Educational psychology studies the nature and process of guidance and counselling.

Study of Statistics. Educational psychology studies mean, median, mode graph, rank difference and co-relation.

A careful analysis of the subject matter of educational psychology reveals that it has three important centres:

1. The Learner. In this area, educational psychology examines the following:

- *(a)* The study of the child, his abilities, aptitudes and aspects of motivation.
- *(b)* Influence of various social agencies like home, school, neighbourhood etc., on the child's presonality.

(c) Individual differences among children.

(d) Various stages of growth and development.

2. The Learning Process. In this area, the following aspects are examined:

(a) Nature of learning and how it takes place.

(b) Theories of learning and how maximum learning in the minimum time period can be achieved.

3. The Learning Situation. Under the aspect educational psychology examines the following:

(a) Nature of classroom management and discipline.

(b) Methods of teaching children with individual differences.

(c) Knowledge about educational and vocational guidance to be given by the teacher to the child.

(d) New techniques of evaluation and their effective use.

(e) Use of techniques and aids which facilitate learning.

Thus, it is evident that the subject-matter of educational psychology is designed to enhance and enrich the learners and to furnish teachers with the knowledge and understanding so as to help them bring improvements in their quality of instruction.

STEPS IN THE PSYCHOLOGICAL EXPERIMENT

Broadly speaking the following steps are involved in a psychological experiment:

1. **Defining the Problem.** First of all the field of study is defined. The experimenter knows clearly which factors to control, which factors to change, what changes to observe and with the help of which material.
2. Selecting the subject or the individual on whom the experiment is to be performed.

3. Instructing the subject or telling him his functions during the experiment.
4. Getting introspective report of the subject. In psychological experiment we have to deal with living human beings and while stating the result of an experiment we must mention the reaction of the subject towards the experiment and his mental state during the experiment.
5. Setting the material for observation.
6. Observing the responses of the subject.
7. Anlaysis of the data obtained and interpretation of the data.

Problems Tackled. The following problems have been tackled in Experimental Psychology:

1. Sensation, sense training and preception.
2. Habits.
3. Laws of learning and transfer of training.
4. Attention, its span and duration.
5. Memory and methods of memorising.
6. Work and fatigue.
7. Intelligence and its measurement.
8. Thinking and reasoning.
9. Imagery and association of ideas.

Criticism. 1. Further, Experimental method has serious limitations when it is applied to Educational Psychology. While dealing with human beings it cannot be as objective as it is claimed by its advocates, because human behaviour cannot be controlled to that extent as is possible in the case of physical and chemical phenomena.

2. Secondly, human behaviour under controlled conditions is different from spontaneous behaviour. Therefore, the experimental findings should be cautiously applied to the

spontaneous behaviour of human beings under normal and free conditions.

Concluding Remarks. It is not always possible for every teacher to experiment scientifically because of no special training in doing so and of lack of facilities for experimentation in schools. Yet he can adopt new teaching methods with better results.

1. New teaching methods suggested by Psychology should be tried.
2. Experiments can also be carried on in the development of character and leadership. Schemes of self-government and of giving responsibility to students can be tactfully introduced in schools.
3. Experimental Psychology helps the teacher in dealing satisfactorily with the backward, the deficient, the emotionally deranged and the delinquent child.

IMPORTANCE AND UTILITY OF EDUCATIONAL PSYCHOLOGY

The contribution of educational psychology to the theory and practice of education is rich and varied. The knowledge of educational psychology is important as it provides teachers with some basic skills and guidelines to solve the problems of teaching-learning process.

Teaching is not everybody's cup of tea *i.e.*, everyone cannot teach.

Kuppuswamy's view, "Psychology contributes to the development of the teacher by providing him with a set of concepts and principles."

According to John Adams, "Teacher should know John as well as Latin". It means teacher should know child and subject-matter. A teacher should know the nature, capacities, likings and aptitudes and attitudes of the child. Child is like a book, teacher should know each and every page of it."

Skinner's view, "The teacher needs psychology to bridge the lives of the young and the aims of education in our democratic society."

Teaching is an art. Knowledge of educational psychology is very useful and indispensable for the teacher because it gives knowledge to the teacher.

Knowledge of Behaviour. Educational psychology assists the teacher in knowing the behaviour of the child at different stages of development. It also helps the teacher in understanding the physiological and psychological basis of behaviour, *i.e.,* nervous system, glands, instincts, emotions, sentiments, motives, play, intelligence, heredity and environment etc.

Knowledge of Innate Nature. The child has got natural urges instincts, potentialities and propensities. These innate qualities are the "Prime movers" of his behaviour. The teacher who knows psychology can make his teaching very sucessful while keeping in view innate nature of the child.

Knowledge of Guidance. It helps the teacher in giving guidance to the pupils by having an understanding of interests, abilities, aptitudes, achievements, problems, educational and vocational plans of the pupils.

Knowledge of Unconscious Mind. It helps the teacher in knowing the unconscious mind of the students and plays very important role in the development of the personality of the individual.

Knowledge about Himself. It helps the teacher to know about himself. He learns the psychology of being a teacher and acquaints himself with the traits of a successful teacher.

Understand the Individual Differences. No two individuals are alike. The teacher with the knowledge of the kind of individual differences may adjust his teaching to the needs and requirements of the class and thus, may be helpful in creating conducive environment in the schools where the students can develop their inherent potentialities to the maximum.

Understand the Nature of Classroom Learning. The knowledge of educational psychology provides a teacher the knowledge of learning process in general and problems of classroom learning in particular. The teacher by the knowledge of educational psychology can understand the principles of learning and various approaches to learning process, problems of learning and their remedial measures and also about factors affecting and guidance for effective learning.

Understand Development Characteristics. Children pass through different stages of development as infancy, childhood and adolescence. These developmental stages have their own characteristics. If the prospective teacher knows the characteristics emerging at different stages of development, he can utilise these characteristics in imparting instruction and moulding their behaviour according to the specified goals of education.

Understand the Problems of Childern. By studying educational psychology a teacher may understand the causes of the problems of the children which occur at different age levels and can successfully solve them.

Knowledge of Mental Health. By studying educational psychology teacher can know various factors which are responsible for mental ill health and maladjustment and can successfully help in mental hygiene.

Understand Effective Teaching Methods. Educational psychology gives us the knowledge of appropriate teaching methods. It helps in developing new strategies of teaching. It also provides us with the knowledge of different approaches evolved to tackle the problems of teaching at different age levels.

Curriculum Construction. Psychological principles are also used in formulating curriculum for different stages. Needs of the students, their developmental characteristics, learning patterns and needs of the society all are to be included in curriculum construction.

Measurement of Learning Outcome. Psychological tools help the teacher to assess the learning outcome of the students and also to evaluate his teaching methods for required modification.

Research. Eductional psychology helps in developing tools and devices for the measurment of various variables which influence the behaviour and performance of students.

Understanding of Group Dynamics. Educational psychology helps the teacher to recognise the importance of social behaviour and group dynamics in classroom teaching learning.

Helps to Develop Positive Attitude. The teacher training programme aims to develop positive attitude towards teaching profession and provides the prospective teachers with the necessary competencies to meet the classroom challenges. Training colleges provide the knowledge of organising the subject matter in a sequential order to suit the needs of the class. The trainees are also aquainted with the techniques of motivating children for learning.

Problem of Discipline. With the knowledge of educational psychology teacher utilises the importance of indirect discipline rather than corporal punishment. It tells the teacher that discipline should be self-discipline, dynamic, positive and constructive through participation in purposeful activity. Pleasure and pain, reward and punishment, praise etc., should be judiciously used. If the teacher is unaware of the principles of educational psychology he may be unable to solve the problems of his students and thereby fail to induce order and discipline among them.

Use of Audio-Visual Aids. Educational psychology has helped the teachers to make use of various types of audio-visual aids in classroom teaching so as to make the concept more clear, definite and learning to last longer.

School and Class Administration. Former autocratic method of administration in school and class has been charged by democratic way of life where in the teachers and administrators are more democratic, co-operative and sympathetic and problems of administration solved by mutual discussion.

Time-Table. The knowledge of psychology is helpful to the teacher in preparing time-table. He should keep in mind the relative importance and toughness of different subjects and level and index of fatigue of the students.

Use of Innovations. Activity-centred teaching, discussion method, microteaching etc., are some innovative ideas adpoted to improve the teaching learning process.

Co-curricular Activities. Activities like debate, drama, games are given due importance alongwith theoretical subjects for the harmonious development of the personality of children.

Production of Textbooks. Educational psychology has helped in planning of textbooks according to the intellectual development of children, their needs and interests at different age levels.

Undoubtedly the study of educational psychology may be very helpful to equip our prospective teachers with necessary skills to deal with classroom teaching learning problems.

REVOLUTION IN EDUCATION INDUCED BY PSYCHOLOGY

The extent to which psychology has produced into the field of teaching and learning education is so impressive that only the term revolution can be applied to it with any accuracy. In pointing to the changes in education caused by psychology. Ryan has written, "In many schools of today one finds an atmosphere of friendliness and happy activity. Much of the traditional formality, the forced silence, the tension, the marching is gone. Childern's voices are heard in the halls and classrooms. The younger children come gaily down the stairways, naturally and relatively unrestrained, the older boys and girls through the corridor or outside walk making their way to school rooms, shops, studios, libraries, laboratories and playing fields to tasks that mean something to them, that make demands upon their energies and their imagination, that often involve hard, difficult work, but work that they recognise as creative. Beauty of surroundings is considered a first, requirement in these schools—there are flowers about, brightly coloured murals painted by the children, attractive informal work rooms for the various groups....Art and music have begun to play the role that belongs to them as fundamental in education and life."

1. Child-Centred Education. In early times the teachers were mainly engaged in imparting various kinds of information to the children. In those times children were less the centre to attention, the school claimed more of their time. Today, it is the child who is the centre of attention in education. Every child is individually considered and treated. Efforts are always made to adapt the curriculum of study to the needs of the child with the result that the brilliant and dull witted children do not study the same subjects. Applied psychology has directed attention to the differing abilities and capacities of people. Such an attitude has made it possible for the teacher to gauge the special abilities of his wards and to make the best arrangements for the guidance of individual children. The teacher of today not only must be acquainted with teaching but also with his students since his profession is now centred about the child and not about the subject or teacher. The consequence of this kind of reasoning is that what matters isn't the beauty, knowledge or ability of the teacher but the extent to which he succeeds in developing the personality of his students.

2. Reform in Curriculum. Psychology has also managed to bring about important variations and reforms in educational curricula. Now it is the mental age and not the chronological age of the child that helps to determine the course of study to which he is to be subjected. Intelligence tests reveal the mental age of the child. Different courses are devised for brilliant and deficient children. After the eighth or delta class, the curriculum is sub-divided into various parts such as literary, scientific, agricultural or art class. The child then has the opportunity of selecting his subject according to his interest and aptitude. Various extra-curricular activities are organised for the emotional and sentimental development of the child. These include picnics, dramatic societies, etc. Other programmes are organised in order to develop the qualities of leadership, public speaking, etc., in the child. But no course of study or programme is foisted upon the child against his inclination. Inspiration is awakened in him according to his interests and his tastes. Education has now been made more interesting and appealing with the use of audio-visual aids such as television, film projection, etc. Many schools are also equipped with radio sets that provide both entertainment and information to the child. The use of analogies to penetrate the defences of the child is an inportant method in education. Childern who choose to study history are also taken to visit historical places. And even a glance at the curriculum of basic education makes clear the extent to which psychology has revolutionised education.

3. Process of Education. Previously, education was believed to be a comprehensive process that trained the individual, provided him with a moral character and made him more cultured. In psychology as it is understood today, the conception of mind has completely changed. Analysis has led to the discovery that many kinds of forces and mental activities take place within the human being so that for the development of the child it is necessary to provide him with different kinds of programmes and subjects so that he may succeed in properly developing all his mental faculties and abilities. At present, variety is believed to be essential not only in education itself but it is equally held to be true that the education of one individual subject includes many different activities and functions. The teacher should be aware of all these functions. It is only than that he can understand the difficulties of the children in understanding the subject. If a child fails to memorise some one lesson, it can hardly be taken to indicate only a lack of effort on his part and no other cause. It is equally probable that he has a weak memory, or is not intelligent, or has no interest in the subject or that he is suffering from ill health. Learning is a complex activity and it is only through a complete knowledge of its various parts that one can diagnose the difficulty of a particular child in memorising or learning a lesson. The teacher has no alternative but to attend to the activity of learning.

4. Discipline. Discipline has an important place in education. One of the aims of education is to instil respect for discipline in the child. But the recent discoveries of psychology have changed the ways and means in which discipline is taught. Previously physical punishment was one of the chief methods of teaching and imparting discipline in the children. The main motive behind the discipline was fear of physical violence and injury. But instead of injuring the body of the child it injured his personality by making it distorted. Psychologists attracted the attention of enlightened people to the injurious effects of physical punishment and also suggested alternative means of making children disciplined. One suggestive example is here described. The inmates of a hostel were given to indiscipline during lunch time. They threw morsels of food at each other and wasted much more. The warden made every effort to stop this nuisance used every means of arousing fear in the children but without success. He never discovered the name of the mischief makers and neither did the boys give up their behaviour. Some time later he was replaced by a new warden who resorted to psychological methods for solving this menace. He made an announcement to the effect that the best behaved table with the most seemly boys would win a flag that would entitle them to twice the normal ration of the most succulent food. The strategy took immediate effect and the indiscipline promptly ceased. The children seated at various tables were now busily engaged in trying to look more disciplined and seemly than all the others so that they now even prevented the more hardened and incorrigible boys among them from mischief. In this way, the warden took advantage of the psychology of children to effortlessly perform a task that had completely foiled all the efforts of the earlier warden with his shock and fear tactics. Nowadays it has become customary in schools to allow the children to adopt discipline rather than foist it upon them. At the college level it is from among the students that the prefects are chosen to create and maintain discipline among their fellow mates and to help the proctor in maintaining order within the college. Very recently there was a high level debate on the spreading indiscipline among university students in India and almost all speakers put forward psychological solutions because the problem of indiscipline is fundamentally and basically a psychological one. It need hardly be pointed out that the more a teacher knows of child psychology, the easier will it be for him to create discipline in the motley group with which he is faced in the class.

5. Individual Differences. In days gone by, the same curriculum was prescribed for all stduents in the class. The development of applied psychology led to the fact that different individuals differ from each other in respect of their interests, intelligence, ability, capabilities, etc. This knowledge of individual differences proved a variable source of revolution for education. Now-a-days the essential pre-requisite to guidance of an educational nature is a knowledge of the subject's ability and interests. Choice of his curriculum and other extra-curricular activities is definitely governed by these individual differences.

6. Education on Personality. The modern methods of education place great emphasis upon the education of personality. Education is now understood to mean something more than mere reading or writing, it is also understood to involve the development of personality. Many schools have resident psychologists who solve problems related to personality of the students and give advice to teachers and parents in this regard. At places there are also child guidance clinics, that make special efforts toward the correction and normal rehabilitation of problem children or juvenile delinquents. A study of psychology is an essential part of the curriculum of teacher's training. In addition to these individual efforts most states and districts have a bureau of psychology whose functions are to solve all problems of students relating to their personality, to give personal guidance and to give advice to their parents and teachers regarding them. The truth of the matter is that the psychologist has become a link for the child's adjustment between the home and the school since such an adjustment helps to improve his personality. In the role of specialist his contirbution to the school has become indispensable.

7. Teaching Method. In the manner already described psychology has changed teaching methods lock stock and barrel. All new research aims at evolving teaching methods that induce the child to learn for himself and thus, himself achieve his development. The means of a changed teaching method are radio, cinema, cultural programmes, debates and other competitions, picnics, visits to historical places, games, election contests, students' unions, etc. It is psychology that has contributed the Kindergarden and Montessori methods of teaching. Speical learning and teaching methods have been evolved for the use of blind, dumb and deaf children and mentally backward or handicapped children.

8. Process of Learning. The experiments that psychology has carried on in connection with the process of learning have led to the discovery of many laws that economise of time and yet produce good results in terms of material learned. One example would be Thorndike's laws. It is the psychologist who has stressed the importance of reward and punishment in learning. Research in psychology has aimed at discovering the efficacy and inefficacy of the various methods of learning. The employment of the various methods of learning is improved and made more scientific by a knowledge of their qualities and drawbacks.

9. Mental Testing and Guidance. Mental testing and guidance are important factors in applied psychology. Psychologists have evolved scientific tests to measure intelligence and other abilities. Students can be guided in educational and vocational matters with the help of these tests. Now, education aims both at the complete development of the individual as well as his best possible adjustment to his vocation and of his vocation to his natural abilities. Teachers need the help of psychologists in either of these two matters. Psychologists solve the problems of students through personal guidance and inform them of the job they are best suited to through vocational guidance. In this way, the co-operation and co-ordination between the teacher and the psychologist in the field of education is on the increase.

10. Extra Curricular Activities. The aim of education is the complete development of the child so that mere book knowledge is insufficient to gain this end. In this connection many other programmes are included in the curriculum on the suggestion of psychologists apart from the normal theoretical knowledge.

11. Reform of Problem Children. One of the most notable contributions of psychology to education is the improvement and reform of juvenile delinquents, absconding children, morons, backward children, unsocial and problem children and advice to their parents regarding them. To this end the state and district psychological bureau assist the educational institutions.

In this manner, educational psychology provides the teacher with the means to attaining knowledge of the student's nature, capacity, characteristics, needs and motive. This knowledge is the equipment for him to modify his teaching methods sufficiently to attain the goal for education.

2

Human Growth and Development

Growth is conceived narrowly as sheer increase in size but in educational and psychological literature the term includes maturation, development and learning. Maturation is the process of coming to maturity or reaching stages of development. The infant matures in the process of becoming an adult; and the parts of the organism also mature. Maturation is a biological phenomenon. It is different from learning which is a psychological fact. Development is change or increase in function due to experience and / or exercise. It is a phenomenon, which consists of progressive changes in the organism. It includes not only changes or alterations in body proportions but changes in function in strength and in co-ordination. Learning is modification of behaviour through experience or practice. It is thus, a particular aspect of development.

The various aspects of development are physical and functional: They are cognitive, social, emotional and moral. Development itself is a unitary concept. The various aspects involved are interrelated. The child, whom we are teaching, is a single being. He is an organic-cognitive-social-moral person. While we can differentiate the physical aspect of his development from the functional one, we cannot separate the cognitive social, moral, emotional and motivational aspects of his development. These aspects overlap each other. They interact with each other.

Conception of Growth

Narrowly conceived, growth is sheer increase in size, height, degree, power, etc. *For example,* growth from conception onwards is a natural thing. It is in this sense used for physical change. As such we speak of muscle growth, brain growth, bone growth and the growth of body in general. The changes produced by growth processes accumulate and are measured directly.

Development is the direction in which growth takes place as a result of factors outside the maturing individual. In this some while growth is a natural process; development is a result of the environmental factors that affect the organism as he grows. So far as his body organs are concerned, the organism grows; but simultaneously there is a change in the expression of his emotion's fear, anger, sympathy, love, apathy, etc. As the individual grows in size or in weight, there is a change in the way he carries his social behaviour. There is a growth in emotional, mental, social aspects, but it cannot be directly measured as we measure growth in physical aspects. The growth in these areas is qualitative in nature and hence we term it differently as development.

In educational and psychological literature growth includes development as it includes maturation and learning. Growth and development are related because both occur together. Hence, there is a tendency on the part of educationists and psychologists to use both interchangingly. Both concepts appear to be synonymous and often are confused. When we speak of physical growth and physical development we are quite clear as regards the difference in the two terms—physical growth is structural and is physiological change that takes place within the individual during the process of maturation. When physical growth takes a certain direction and is facilitated or checked by the environmental factors, which stimulate the growing individual, it becomes physical development. The two terms growth and development are used one after the other in psychological literature; *for example,* we speak of reading readiness as a stage of growth and development at which reading instruction will probably be effective and before which instruction is relatively fruitless. We speak of readiness as a stage of growth and development not only because there is a change which can be directly measured as in mental age and sensory perception needed for reading but also there is a change in emotional control, in desire to read and in experience which cannot be measured.

The task of applied educational psychology is to point out ways in which growth is facilitated, or the ways in which development takes place. The task of the educator is to provide conditions that are favourable to growth : for education is a growth process. The task of the educational psychologist is to discover and create those conditions, which tend to facilitate growth of the individual child. The duty of the teacher is to work in accordance with growth principles rather than against them.

Principles of Growth and Development

The teacher should clearly understand the nature of growth and its principles before he launches upon the task of educating children. These principles are:

(1) Growth and development result from an interaction of Heredity and Environment.

(2) Growth proceeds more rapidly in earlier years.

(3) Growth is continuous and gradual.

(4) Growth moves from Generalisation to Differentiation and Integration.

(5) Growth is a regular step-by-step process.

(6) Each child and adolescent has his own rate of growth.

(7) Maturation influences the results of training.

These principles will be discussed in detail below because they have a far-reaching impact upon principles and practices of education.

1. Growth as a Functions of Heredity and Environment. The development of an organism is a product of two interdependent factors 'what he is' and 'what he has'. 'What an organism is' is determined by his hereditary potential and 'what he has' is fixed by his environment. Both interact on the individual and growth takes place.

2. Growth is Rapid in Earlier Years. In early childhood any structures and functions grow at a much faster rate and as the child grows, the rate of growth slows down. The rate of growth is then negatively accelerated as shown by general forms of growth curves for size, weight, intellect, etc. (see Fig.)

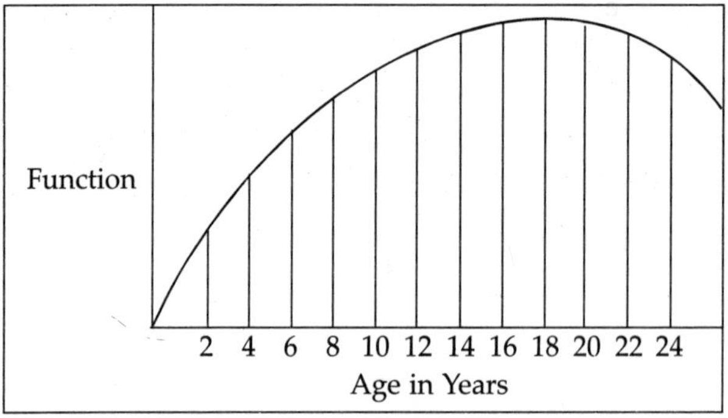

3. Growth is Continuous and Gradual. Growth or development does not occur in spurts. We may talk of stages of development or of growth as infancy, childhood, adolescence but it does not mean that when one stage stops and the other starts. There are no breaks in the process of development.

4. Growth is a Regular Step-by-step Process. Growth is never haphazard. There is order and sequence. *For example,* the baby creeps before the crawls and he crawls before he walks and the child walks before he runs and he runs before he dances. What is true of physical growth is true of every mental process. *For example,* the child babbles before he utters words; he utters single words before he speaks in sentences;

he speaks before he writes; he writes simple: stories before he creates a work of art.

5. Each Child has his Unique Rate of Growth. Some children grow rapidly. Some are slow. Some travel by an ox cart, some by a car and some by a jet plane. Some become prominent when they are still children, some, when they are adolescents and, some never. The rate of growth is not only unique for an individual child for a particular trait, but the rates of growth for all different traits of the child are unique. One child may grow rapidly in one trait in early years and then may seem to have reached a plateau. The other child may show a regular and gradual pattern of growth. The third child may experience a slow rate of growth in early years and then show a spurt, which carries him beyond what is regarded as an average for all.

6. Maturation Influences Results of Training. The role of maturation is illustrated in teaching reading skills. Unless the child is mature enough, he cannot be taught reading.

7. Traits are Correlated. As growth and development, of traits takes place, they grow in a correlated manner. *For example* a gifted child grows physically, socially, emotionally at the same; rate as he grows mentally.

8. Generalisation, Differentiation and Integration in Growth Process. As a child develops in a particular trait, he shows generalised behaviour. *For example,* in the post-natal period when reflexes have developed, a pin-prick causes the whole body to make a movement, but when the child grows he can identify the point where the pin is pricked. Differentiation follows generalisation.

When the baby is shown an attractive toy it tries to catch it with gross muscles, but when he grows older he is able to pick up the object with his thumb and forefinger. There is differentiation. Integration means that all parts are work in harmoniously towards the goal and it takes place concurrently with differentiation.

Growth as a Functions of Heredity and Environment

Growth and development depend upon the hereditary and environmental factors so much that the growing individual is regarded as a product of both. The growing child's behaviour is influenced by his genetic structure, physical body build his rate of maturation. It is also affected by his cultural mileau- the values, ideals, goals and the specific situations, the typical or general environment of home, school and society.

Most educational psychologists now believe in the interactive process of heredity and environment. Heredity provides the potential and environment acts upon this potential and produces what the child is. *For example,* intellectual growth and development are the results of the interaction of the two. No psychologist now believes that I.Q. remains constant; that is, the rate of mental development for a given person can change under certain conditions. For a group of persons it is believed that rate of mental development possibly cannot he extensively changed by providing a stimulating environment. But, however, beneficial and stimulating an environment may be, the change in the rate of mental development is limited by hereditary potential. The less stimulating environment causes a child with high hereditary potential to develop to such an extent as a more stimulating environment can do to the same child. That is why identical twins reared apart show a difference in I.Q. When two children with different hereditary potentials are given equally stimulating environment, they do not develop equally, meaning thereby that even the most stimulating environment cannot affect growth beyond the hereditary potential.

Educational Implications of Heredity Environment

The teacher has to start with the student where he is. The student has come to school with some hereditary potential and it is the teacher who is to be charged with the responsibility of realisation of his potential to the maximum.

Each child has differential amount of hereditary potential; hence the teacher will have to first assess that potential. For this he may use what are known as intelligence tests, achievement test batteries, interest inventories, readiness scales. If he can wisely use the data obtained from such tests about each child he may start well and well begun is half done. How disappointing it is to note that in spite of the importance of such tests, teachers in our schools do not know anything about them. If some Psychological Bureaus have tried to break the ice and have taken tests, the data obtained are simply filed way.

As growth is the result of interaction of the organism with the environment, it means that the teacher should help each child to achieve the greatest possible realisation of his potential. His help may be in the form of encouragement, praise and direction. Every child should be encouraged to do more and to read more, to write more and to perform more intricate tasks and solve more difficult problems. Each child should be praised for good work and given proper guidance. The more stimulating the environment, the greater will be the realisation of hereditary potential. *For example,* if the child meets better birth conditions, gets more immediate and intimate mothering, faces little emotional stress in the home, gets better schools and is taught by more appropriate teaching techniques, he will not only get rapid and vastly increasing knowledge, but will develop emotionally, socially and morally in more desirable directions.

Child's work even when it is below average should not be deprecated. He should be given an experience of success in what he does. The teacher's duty is to direct him to areas where he can get more success. In short, the teacher has to adapt learning experiences to the needs of individual pupil so that he may realise his hereditary potential to the maximum.

The relative influence of heredity and environment on the all round development of the child has been a disputed concern of biologists, educationists and psychologists. But can heredity be sharply differentiated from environment? Are they not influencing the child co-operatively and concurrently? Can the single cell, which is at first the fertilised ovum and later the entire organism, grow without being influenced by both its pre-natal and post-natal environment?

The course of development is influenced by what the organism is (his inherited potential) and by what he has (his environment). Heredity and environment are welded together as interdependent influences by the way one reacts to what he is and has. His reaction to them determines the aspect of development.

Four things are clear now :

(1) The organism gets some hereditary potential. (What he is.)

(2) The environment interacts with hereditary potential.

(3) It interacts with environment. (What he has.)

(4) As a result the organism increases, expands and grows.

Educational Significance of Early Rapid Growth

The pre-school period is characterised by rapid growth. The infant, from 2 to 5 years of age, shows tremendous improvement in co-ordination, perception, differentiation of behaviour and acquisition of skills. The growth in this period is so rapid that you can tell a difference between a 2-year old and 3-year old baby, but it will be difficult to perceive a year when the child grows into adolescence. There is faster rate of language development; there is greater speed of getting along with his brothers, sisters and adults; there is quicker crystallisation of fundamental personality traits. All these developments occur within a short span of four years.

The fact that growth takes place more rapidly during the early years does no mean that it stops after a certain age. The child is continually growing. Even when broad features of personality traits become nearly established during early years, they are subject to modifications later. The idea that in pre-school period broad fundamental patterns of behaviour are determined once for all and no modification needs be done is erroneous. The primary school teacher who is entrusted with educating the child should not feel that his efforts will bring no fruit for much has taken place in the child's life before he enters the classroom at 6+. He should recognise his responsibility to mare the first school experiences as happy as he can.

It is not enough to teach the child reading, writing and arithmetic skills. Acquisition of academic skills such as these, is useful and desirable but more fruitful are the emotional adjustment and social development. The child, who does not meet his social and emotional needs in the primary school, cannot learn academic subjects efficiently. An adequately trained teacher feels the need of making the first few years of school life happy and enjoyable and practises what be feels.

There is rapid growth in the primary school life though not so rapid as in pre-school life. This does not mean that schooling should begin at an earlier age or social experiences should be given in greater amount at the age of 4 or 5. An immature child should not be fitted into a school system. Individual differences must be realised. It means that even if a child is 6+, he should not be forced to enter the classroom if he is not mature enough.

During the early school years there is rapid growth. Though the child is quite active yet he gets fatigued soon either because his body tissues have not become so mature or because rapid growth requires more expenditure of energy. Fatigue causes loss of appetite, restlessness and inattentiveness. The child of Class I, becomes restless or inattentive earlier than that of Class VI. The primary school teacher should shorten the length of periods. A child, who is growing at a much faster rate and spending much energy, suffers from fatigue so much that he does not learn as well as the one who is not growing so fast. The result is that he becomes backward and a slow learner. Primary school teachers should recognise this phenomenon and therefore provide a balanced programme of rest and activity in the school.

Individual Growth Rates and Teaching Practice

The principle that each individual has his own rate of growth asserts that some children grow rapidly and others grow slowly. Growth rates for mental, physical, emotional and social traits are different for different children. *For example,* the rate of mental maturity (Mental Age) is different for different children. What is true for mental age is true for other ages. Besides, particular child has many ages: mental age, social age, emotional age, organismic age and so on. Individual growth rates, however, remain constant. A child, who is above average in intelligence when he is 6+, will remain so when he is 15 +. A child who is tall for his age at 3 will remain so when he is an adult.

The phenomenon that each individual child has his unique rate of growth so far as a particular trait is concerned implies that teachers should deal with the prublem of individual differences effectively. Each child should be provided with the type of training and education best suited to him as an individual and to his expected needs as an adult. An effective teacher realises the nature and extent of individual differences. He recognises the fact that there are individual differences in learning ability, in temperaments, in needs, in interests. in background experiences, in home conditions. in ideals and in attitudes. He adjusts his instruction to these differences.

A good teacher provides a variety of activities to his class. He groups pupils of similar interests and mental development. He schedules his classes of varied difficulty. He tearches out materials fit for different levels pf reading readiness.

A bad teacher is contented with giving the same dose to all and sundry. Hence, there are failures, dropouts, frustrations and poor results.

Physical Development of the Learner

Physical development is studied horizontally and longitudinally. Horizontal studies gives us data regarding growth patterns of children at specific stages of development. Such studies do not give us information about individual growth patterns. Longitudinal studies are progressive studies of an individual and can be used to compare him with himself at various stages.

Different stages that a child passes through from birth to 16 years of age are the infancy (on the average, from birth to 5 or 6 years of age, during which the sensory channels begin to function and the child learns to creep, to walk and speak), the childhood (on the average from 5 to 12 years during which permanent teeth appear and the child learns to read, write and care for himself), the adolescence on the average from 12 years to 18 (in some cases to 20 or 25 during which the development of sex organs bring about the appearance of physical sex characteristics'.

The specific growth changes from birth to adolescence occur with regard to height and weight, body proportions and growth or internal organs.

During infancy a normal child tends to follow certain definite patterns of growth. Height and weight increase tremendously. The infant boy grows on the average 7"' by the age of 9 months and the infant girl grows to 6 3/4" by that age in height. Weight is doubled during the first 6 months. During childhood and adolescence the height goes on increasing. Hereditary background and environmental factors combined determine height-weight ratio for a particular child. *For example,* children of short and stoct parents are generally short and stocky those of the tall and slender are tall and slender.

Different component parts of the skeleton do not grow at a regular rate. When the child is born, the size of bones of arms and legs is smaller than that of the skull. As he grows, the head and torso become proportionally shorter than before and arms and legs become longer. At birth there are no teeth. But as the child grows he gets baby teeth and then a second set of teeth which serve him all his life.

From birth to adolescence all the body systems have different rates of growth. At birth the nervous system is developed sufficiently well. The circulatory system has a balanced functioning too. But with growing years, the heart grows in size and volume. The lymphatic system becomes active at birth and increases rapidly after 12 years of age. The genital system develops slowly during the childhood but grows rapidly from 11 or 12 years of age onwards.

Environment plays a significant role in physical growth and development. Favourable environmental conditions, planned recreation and rest, proper diet, well-ventilated living quarters, freedom from care and anxiety are beneficial for proper physical growth. Better nutrition, better shelter, more adequate life necessities are conducive to better health. Behaviour problems in our children spring up because of malnutrition, unhealthy home environment, lack of rest and recreation anxiety and trouble at home.

The individual is first and foremost a physical being. Therefore, any programme of his education revolves round his physical growth and development. The primary aim of education would then be the development of excellent body, good health, active neuimuscular system.

The knowledge of characteristics of physical growth and development is very essential to the teacher. He should not only be conversant with subject he teaches but he should also study the growth characteristics. The learning equipment, the curriculum, the techniques of teaching have to be geared to the proper growth of the child. He has to be careful of the normal physical growth of the child as he has to be careful of his normal intellectual growth. The child who seems to be growing beyond his years as well as the child who seems to be growing shorty, need his special attention. Even a slight physical abnormality during the childhood creates trouble to the individual. Attitudes, interests and emotional behaviour grow out of physical status. Above normal height or weight or uneven physical growth rate, produce emotional strain and stress in the child. The teacher has to help such a child in solving emotional difficulties and problems.

Many children come from poor homes. Their parents are illiterate and do not know how to take care of them Children are either mal or undernourished. They lack adequate shelter. Living conditions are far from satisfactory in villages and in some urban areas too. Proper attention has to he given to the nutrition of the average school child. A mid-day meal or some other programme with state or public aid may be launched and the problem of maturation can be solved.

Every school has a responsibility for preparing able-bodied youths to serve the nation. Sanitary conditions of school premises, a provision of well-ventilated school rooms, sufficient and well-planned programme of sports and games, periodical medical examination and dental care, are a must for a good school.

GROWTH OF HUMAN CHANGES

As the young person approaches adulthood certain factors in his experience conspire to make him more consciously concerned about himself as a person. He becomes self-conscious in a way that he never was before. The glandular changes described above initiate feelings and emotional qualities that are new and strange to him. The physical changes that are rapidly taking place in body dimensions, in physiognomy, in genital organs, the beginning of the boy's beard growth and voice change and the swelling of the girl's breasts, all contribute to feelings of strangeness. And when the child almost suddenly, as it seems to him, finds that his reach is longer, his feet bump into things like they never did before, that things about him more than ever before get in his way and get knocked over, he begins to feel as if he is dealing with a different person than his former self which he was able to manage so expertly.

The early part of the adolescent period has sometimes been called the "awkward age." Many adolescents are self-conscious concerning their posture and gait. Some show a certain inhibition or hesitancy of movement as though they were holding their muscles in check and did not feel free to let themselves go. Self-consciousness with respect to height may express itself by sloping shoulders and the short adolescent boy may try to compensate for his shortness by an upward tilt of his head, squaring of his shoulders and a tendency to let his weight rest more on his toes than on his heels. Sometimes, in his bearing and by the character of his physical movement, the adolescent calls attention to physical features which he would actually like to conceal.

If there is such a thing as an awkward age, it is during this relatively brief period of puberal change. All of this strangeness, confusion and lack of self-management contributes greatly to what is called the identity crisis. It is he problem of reconciling present experiences, perplexities and embarrassments with ones past and of orienting oneself toward a meaningful future.

There are, of course, wide individual differences in the seriousness and the disturbing aspects of this identity problem. Many young people appear not to be bothered in the least. They continue to be their outgoing selves, maintaining their usual easy relations with their peers, older persons and with parents and family. Others, however, tend to "retreat into themselves," becoming affected or difficult to live with or critical, antagonistic and alienated from parents and adults generally. These youngsters become strongly peer oriented, but at the same time they may experience difficulty in their peer relations as well. Such individual differences would be difficult to trace to their origin. Differences in congenital temperamental predispositions undoubtedly are often important factors. And, as Erikson (1968) suggests, experiences in early parent-child interaction, which contribute or do not contribute to a sense of trust in the environment, such as the care of a mothering person during early infancy who is "warmly dependable," may also be at the basis of a difficult identity crisis.

Adolescent Sensitivity to Deviations in Physical Traits

As was noted in young people, although they may want to be different to a degree to express their individuality-also have a strong need not to deviate too much physically from the currently popular stereotype. Girls entering adolescence, as a rule, particularly if their inherent body type tends to be in the direction of endomorphifan, react quite negatively to a condition of overweight. Neither do they want to be too tall. Boys, on the other hand, are likely to be disturbed by an unusually short stature, They are especially sensitive to and disturbed by signs of sexual immaturity due to the late onset of pubescence.

This latter problem was noted particularly in a pair of brothers, ages 12 and 14, whose development was followed longitudinally. The older of the two entered the pubescent phase somewhat later than average, while his younger brother tended to be an early developer. As the older boy approached his fourteenth birthday he still maintained all the characteristics of a prepuberal child—a relatively short stature, a high-pitched voice and no obvious signs of genital organ growth. At the same time his 12-year-old brother was rapidly exhibiting all the signs of puberty. For a period of several months this contrast in maturational status was a source of considerable anguish on the part of the older sibling. In an interview he expressed much concern as to whether he was "ever going to grow up." This disturbance was reflected in his relations with his family, particularly with his brother and his age peers as well.

H. E. Jones (1949) in his studies of adolescence has pointed out that the effects of early or late maturing often affects boys and girls differently. Girls who mature earlier than average often experience it as a disadvantage in their relations with peers of both sexes. An early-maturing adolescent girl is much ahead of boys of her own age in size and in intellectual and social development. Normally adolescent boys and girls, even when they are of the same age, differ as much as a year or more in these areas of development. Such sex differences are often sources of some difficulty on the part of school personnel in dealing with mixed-sex groups of same-age youngsters.

In contrast with the early-maturing girl, the boy on an early-maturing schedule may experience some advantages. He will be larger and stronger than many of his age peers. Jones and Bayley (1950) found that early maturing boys were more attractive physically, less affected and more relaxed when their behaviour was judged in comparison with non-early maturers.

Young people in early adolescence often are particularly sensitive to and concerned about relatively trivial deviations from the usual, or what they regard as desirable, in physical make-up. Such things as too slender or too plump legs, breasts that are underdeveloped or too large and pendulous in girls and in boys, what they regard as poor muscle development can be matters of keen personal dissatisfaction. Bow legs and calves that are not sufficiently round are sometimes causes of anguish in either sex, especially during times when clothing styles call special attention to such irregularities.

Menstruation and its Psychological Impact

The menarche—the beginning of the menstrual cycle of a young girl—is the most reliable indication that she has reached puberty and is now capable of conceiving and bearing children. There is no strictly comparable criterion of sexual maturity in boys.

Menstruation is a natural phenomenon, a normal phase in the cyclic production and discharge of mature ova. In the past there have been many misconceptions and superstitions about this very important normal function. It has been referred to as a "monthly sickness" with which women are afflicted. Such views are not now widely accepted.

However, the fact that menstruation, which involves rather complicated bodily changes largely physiological in nature, begins at the onset of pubescence can have an important psychological impact, according to several studies . It is true that the process is frequently accompanied by unpleasant physical symptoms such as abdominal cramps, headache, backache, nausea and fatigue. The attention of the pubescent girl is naturally drawn to this cyclical event in its association with puberty and she may under certain circumstances regard it with resentment as a penalty for being a female. It is also true that even now in our culture menstruation is sometimes associated with considerable shame, embarrassment, secrecy and dread. It is, however, accepted unemotionally and even eagerly by the majority of girls.

As was noted earlier, the attainment of sexual maturity, or the ability to reproduce, in boys has no single indicator criterion which compares with the menarche in girls for 'reliability. The full growth of genitalia and ejaculation of semen are perhaps the most definite indicators. These and other changes and events marking the developmental phase also have their potent psychological impacts. Voice change, clearly a result of structural maturation (a qualitative change in the vocal organs) is perhaps most embarrassing to the boy. The unpredictable "breaking" of the voice in social conversation is very disconcerting to some boys while to others it is nothing more than a slight annoyance.

Motor Functioning

We have just been concerned with the widespread and dramatic physical and morphological changes which characterize the approximately to-year period which we call adolescence. As we now center our attention more specifically upon the functional aspects of that development we shall note many equally interesting changes—changes in strength, speed and smoothness of performance, as well as in the importance of motor abilities and skills in the adjustments of individual youngsters. We shall particularly note some important sex differences in the nature and direction of change in interests as well as actual abilities in areas of physical activity. Wide individual differences among both boys and girls in interest and ability to perform are seen in relation to school programs in physical education.

Changes in Motor Performance

The rapid growth of the body during the puberal phase and the great amount of physical activity that continues to characterize adolescence, result in marked increases in physical strength. The youngster normally very soon becomes accustomed to his new and changing physical dimensions. Precision and smoothness in motor performance, as a rule, are soon developed. The adolescent's senses are keen and his reaction time is relatively short. Consequently, his speed of performance is generally high. Unfortunately, however, due to his lack of experience, his judgment and discretion often do not match the level of his behavioural pace. He, of course, carries over his pre-adolescent interest in sports and with his added size, strength and speed of reaction his athletic skills generally are perfected. Thus, we see a marked change from the so-called awkwardness of early pubescence.

These enhanced athletic abilities often play a significant role in social relations, particularly among boys. Comparable developments in girls, however, may play an equally important but more subtle role in the social development of adolescent girls. Clearly, nothing lends more to a boy's sense of importance and status during his high school years than recognized competence in athletics. As everyone knows, the football star is the envy of his male age peers and is often the idol of the female crowd.

Cognitive Development

From an earlier discussion it will be recalled that the course and progress of brain growth from birth to adulthood closely parallels that of intellectual development. It is only logical to assume that in the general speed up of body development, the puberal growth spurt described earlier, the brain would be involved. Students of cognitive development are in general agreement that a significant rise in quality of mental functioning also occurs during approximately the same period of time as the physical growth acceleration.

Influence of Piaget

We reviewed briefly Piaget's account of the development of mentality, or cognitive functioning. Cognitive development, as we have seen, begins with what is probably a complete lack of mental functioning at the time of birth and culminates in adult capacity to function with understanding, effectiveness and facility in relation to the external environment. To reason, to solve problems through the study and manipulation of things, events and relations not perceptually present is perhaps the most complicated and involved aspect of human development. It is an area in which much research has been done and is now underway.

Perhaps no other person has stimulated as much research or has directly contributed as much to the present level of understanding in this area as the Swiss psychologist Jean Piaget. His meticulous observations of child behaviour and his many "experiments" have led to the formulation of a complete and integrated theoretical account of cognitive development. Many other highly competent psychologists have made important contributions. Many of them have raised questions about Piaget's findings and theoretical statements and many disagree with him at points. Piaget's account, nevertheless, continues in its general outline, to be widely accepted. We shall now briefly review the steps in the development of intellect during the period of adolescence as traced by Piaget and his co-workers.

Emergence of the Adult Level of Cognitive Functioning

According to Piagetian thinking, at about 11 or 12 years the youngster normally begins "freeing himself from the concrete" and orienting himself toward the non-present and the future (Piaget and Inhelder, 1969). At this level, however, he still thinks in terms of "concrete operations." Concrete operations provide a transition between schemata of action—the semiotic function and the general logical structures, the "formal operations" of the mature intellect.

The child at this concrete operations level thinks in terms of mental representations. He is no longer limited to representing by "imitating" an object or even in the form of an overt act. The representation is "decentered" from action. His operations, that is, his thinking, are concrete, however, in the sense that they are tied to actual perceptual experience. He cannot yet deal mentally with propositions, with hypothetical situations not directly experienced perceptually. This level of thinking is exemplified in the simple "conservation" experiment in which all of the water in a glass container is poured, before the child's eyes, into another glass, which is taller and more slender. If the child is at the concrete operations level, he will say that, "it is the same water," "it is the same amount." He can represent mentally the attribute of a constant amount because he has perceived the dimensional changes in the water mass.

The transition from this stage to that of formal operations comes about gradually.

Formal Operations

The advance the young person makes in the quality level of his thinking as he frees himself from the "concrete"—from representations of things and of their perceived qualities and relations which he sees as true—and begins to reason correctly "about propositions he does not believe, or at least not yet; that is, propositions that he considers pure hypotheses" (Piaget and Inhelder, p. 132), is indeed remarkable. And, as was noted earlier, this step forward is quite contemporaneous with the generall pubescent acceleration of organic development.

The final "decentering" process in which one frees oneself for "propositional" thinking, for the consideration of all possible ways of solving a problem, requires time. It is not until about age 14 or 15, on the average, that the full level of formal operations is attained. The individual by that time can isolate the various elements of a problem and systematically explore a variety of approaches to its solution. The formal operations level of thinking thus, makes possible the adoption of a generalized orientation to problem-solving. This may involve the controlled, critical evaluation of hypotheses, with final selection and testing-the thinking of the scientist.

Emotional Adjustments of Adolescence

According to Piagetian theory, this continuous yet sequential development of the cognitive functions is paralleled by an equally important emotional growth and increased capacity for social functioning. Alongwith the repeated process of "decentering" from earlier modes of behaviour toward more "mature" levels of cognitive functioning, this increase in capacity for higher level emotional and social functioning, can also be seen. *For example,* there is a gradual growth from complete egocentricity through varying degrees of capacity for sensitivity and empathy with respect to the feelings and welfare of others. There might also be growth in the capacity to defer gratification and to control and regulate the nature of one's emotional expressions with due regard to the situation and the presence of others.

Special note was made earlier of the psychological impact of growth and maturational changes of the puberal period. The changing pattern of endocrine glandular secretions, bringing about changes in the chemical content of the blood, result in new patterns of physiological functioning. These changes naturally affect directly the emotional experiences of the young individual. They give rise to the so-called sex drive and the whole related area of interest in and feelings toward members of the opposite sex.

Sex Differences—Changes with Development

During middle childhood there is usually little evidence of interest on the part of either boys or girls in the opposite sex. On the contrary, a kind of sex antagonism, real or pseudo, often builds up. Boys play with boys and organize their "no-girls-allowed" play clubs. Girls likewise prefer girls as companions. However, the situation begins to change in late preadolescence. Boys' interest in girls begins to develop. There are wide differences among boys as to the extent to which they manifest this interest, depending very largely upon their experiential backgrounds in family, school and social groups. Some boys strongly deny and try to hide any interest they might have in girls. Other boys feel no need to hide their interest and they begin to enjoy friendships and association with girls. Prepuberal girls are more likely to frankly show their interest, especially in boys' activities. They more frequently participate, when allowed, in the rough-and tumble games and stunts of the boys.

With puberal change, as already noted, come changes in emotionality in general, with a strong surge in interest in and need to associate with the opposite sex. At first this interest is likely to be quite general and nonspecific. There is simply pleasure in associating, talking together, bragging, "showing off," trying to impress one another. It is in such group associations, however, that the more intense mutual pair attachments develop. Pairing off and "going steady" tends to begin. However, by this time girls are one to two years ahead of boys in physical development. This results in a general age differential between boys and girls as they begin these more intimate relationships. Girls, who are also ahead of boys in social development, see the boys of their own age as crude and childish and they turn their interest to older boys who are more nearly on their level developmentally. The "rejected" boys, the age peers of these girls, do not feel rejected. They are usually rather disdainful of the "put-on" airs of their former girl associates. A year or two later these boys begin to relate to girls somewhat younger than themselves.

Nature of Adolescent Sex Adjustment Problems

It is, of course, at the point when boys and girls begin to pair off and to establish the more intimate, one-to-one relationships that problems are likely to begin.

Depending on one's individual make-up and background of experience, any of a number of problems might arise when person-to-person intimacy begins. At first there may be simply the problem of just what does one do, how does one behave when alone with one of the other sex. Relationships established early, of course, are not all of the same nature. In some situations a rather serious problem of intelligently dealing with one's urgent "need" for sexual gratification develops. How this problem is dealt with, together with whether any inner conflict develops between what one does and one's internalized standards, depends upon factors in one's background.

As was suggested earlier, in the beginning of mixed-sex group associations the person-to-person relationships are generally of a very nonspecific nature. That is, the pleasure of associating together is largely based upon "differing social roles and complementary psychological natures." Boys and girls enjoy being together because they are of different sex and thus, complement one another in the group situation. According to some students of adolescent behaviour, this same kind of interaction is sometimes carried over into the later, more intimate pair relationships (Staton, 1963). The couple find that they share many common interests and a mutual respect and a closeness develops between them.

There are, of course, the more highly emotionally charged pair relationships. According to some observers (Grant, 1957;

McKinney, 1960), these are of two general sorts, differing mainly in the quality of the pre-dominant pattern of their emotional interactions and motives—the "amorous" or love attachment and the more specifically erotic or "genital" kind.

According to this view, heterosexual attraction may be based upon motives other than the specific desire for sexual intimacy. The amorous type of relationship presumably is one in which there is genuine and complete devotion of boy and girl to each other. There is sensitivity to each other's feelings, desires and aspirations and effort is made to facilitate each other m his or her personal fulfillment. There are mutual feelings of closeness and spontaneity in expressions of love and affection.

Not everybody, according to recent research, is capable of this kind of love experience. The findings suggest that such a temperament, that is, the capacity to give and to receive love, has its basis in sensory and affective experiences of early infancy. The work of a number of investigators indicates that the adequate meeting of the infant's need for appropriate stimulation and maternal care is the foundation for the development of affectional capacity.

The "genital" type of relationship, by contrast, is one in which the primary bond is the realization, or the promise, of mutual gratification of erotic love. There would be other expressions of regard for each other and other kinds of intimacy, but erotic gratification is the primary motive.

Sex Differences in Sexuality

So far our discussions of adolescent adjustment problems have proceeded with the 'implicit assumption that there is much similarity and mutuality between the sexes in manner and mode of coping with common problems. And indeed there is. But there are, nevertheless, certain problems and difficulties that are peculiar to the boy or the girl. Some of these, of course, are associated with a commonly shared area of difficulty; others are quite distinct and specific to either boys or girls.

With respect to the "sex drive," as such, *for example*, there are some rather significant sex differences. The main difference here seems to be not so much in the overall strength of the drive as in its immediacy and readiness of arousal. At any time and under almost any set of circumstances erotic feelings and interests may arise immediately in the boy's consciousness in response to visual, auditory, or other sensory stimuli. This aspect of the drive in boys is "imperious and biologically specific." The boy also seems subject in some degree to direct 'hormonal stimulation. Feelings thus, aroused give rise to erotically charged thoughts and imagination.

In girls, on the other hand, the early "sex drive" is not specifically erotic in nature. The girl is likely to experience, instead, rather "disturbing longings, unknown stirring of feelings, feelings of tenderness and soft affection, a desire to love someone ... " (Staton, 1963, p. 351). Whereas boys readily become strongly aroused sexually through nothing more than thinking or conversation about girls or viewing female nude pictures, girls usually require much more direct stimulation of the erotogenic zones (lips, breasts, genitalia). "Their sexual feelings, too, are enjoyable for their own sake, constitute a pleasurable state of feeling to be maintained," rather than, as in boys, a feeling which immediately becomes erotic, a state of high tension which tends to mount, propels toward climax after which it rapidly subsides.

Individual Difference

There is, of course, wide individual variation in both boys and girls in the nature and strength of sexual motivation and expression. The evidence suggests, however, that the range of individual differences in this area is greater for girls than for boys. Boys more frequently and more closely follow the "typical" male pattern. Girls, on the other hand, show a wider spread of variation. A greater proportion of girls, at the older adolescent ages, in both "readiness" to function and in mode of sexual response tend to approach the pattern that is more characteristic of the male.

Sometimes from the dawning of physical signs of adolescence and occasionally even before, a boy or girl will display unmistakable signs of erotic sexual desire and make physical approaches, seductive in the case of girls and aggressive in the case of boys, aimed at sexual activity.

Cultural Influences

Not only is the young person beset by compelling new physiologically based drives from within but he also finds himself on the threshold of a complex, inconsistent and confusing adult world on the outside-a world with which he has had little involvement or concern heretofore. In his earlier, relatively carefree years of middle childhood he was not immediately faced with the problem of relating himself to and becoming a member of adult society. He was not yet concerned very seriously with his probable future status or his realm of functioning in it. His primary concern very largely had been his relations with his peers and his performance status among them. Because of this preoccupation with things that were of concern for him as a child, along with the fact that he received relatively little attention from adults, he finds himself on that threshold knowing little about the real day-to-day activities of adults, their concerns, their thinking. In most instances he has probably seen more of the sordid and the frivolous aspects of adult life, the recreational, pleasure-seeking activities. He is also likely to have observed and otherwise been made more aware of the competitiveness, the deceitfulness and the lack of trust-worthiness in society than of the truly social concerns, the honesty and integrity, that characterize most adults. In short he is not prepared for the adjustment problems he faces.

Self-identity Problem

Youth has always been a problem, which has concerned adult society. This concern, however, has been not so much with the inner perplexities and difficulties young persons themselves experience as with how to manage them, how to get them integrated with a minimum of disruption.

Of course, the young adolescent has his own problems. He almost suddenly finds himself with the bodily dimensions of the adults around him and with his adult like mental capacity he senses what the expectations of society with respect to him will be. Yet at the same time his parents and others from whom he might expect to get most help probably continue to talk to him and generally to treat him as a child. Before he can do anything else he must understand these and many other contradictions.

This task is self-definition. Adolescence is the period during which a young person learns who he is and what he really feels. It is the time during which he differentiates himself from his culture, though on the culture's terms. It is the age at which, by becoming a person in his own right, he becomes capable of deeply felt relationships to other individuals perceived clearly as such.

Thus, in the later school years young people, beset with physiological revolution of their genital maturation and the uncertainty of the adult roles ahead, seem much concerned with faddish attempts at establishing an adolescent subculture with what looks like a final rather than a transitory or, in fact, initial identity formation. They are sometimes morbidly, often curiously, pre-occupied with what they appear to be in the eyes of others as compared with what they feel they are and with the question of how to connect the roles and tasks cultivated earlier with the ideal prototype of the day. In their search for a new sense of continuity and sameness, which must' now include sexual maturity, some adolescents have to come to grips again with crises of earlier years before they can install lasting idols and ideals as guardians of a final identity.

Erikson thus, sees the achievement of a sense of identity not so much as the meeting of a sudden, full-blown "crisis" related only to the period of adolescence but rather as a long-term developmental process having its beginning in the earliest experiences of infancy, with "stages" in its epigenesis along the way. This crucial adolescent stage in the process normally results in a rather thorough revision of the self-concept.

As Erikson points out, an important factor determining the "satisfactory" working through of this problem is the adequacy with which one has been allowed to pass through each of the earlier critical periods of childhood and to achieve the related developmental tasks. It should be realized, too, that this universal problem and responsibility of discovering one's true self and of working toward the fulfillment of one's potentials, really continues throughout life.

An important facet of adolescent change, certain aspects of which we have already noted, is an upsurge in the so-called sex drive. One facet of normal erotic expression is the experience of "tenderness"—of loving and cherishing another individual as a person. As was noted earlier, this quality of tenderness seems to be more characteristic of girls than of boys. It will be recalled also that much emphasis has been placed upon the importance of early infancy as a time when the foundation is laid for the capacity to give as well as to receive affection and love. A wide range of differences is evident in individual level of this capacity. However, along with the adolescent advance in mental capacity and associated with. the simultaneous upsurge in erotic functional capacity, is growth in the capacity to love. The adolescent, then, must deal with this new urge to erotic love as an attribute of himself in his task of coming to know himself.

Difficulties

Reference to the ego-identity problem suggests a single common crisis of adolescence. The problem is, of course, a common one, in the sense that certain of its broad aspects are essentially the same for all. But since every person at the beginning of adolescence, or at any other level of development, is an individual, different from every other, the particular identity "crisis" he faces is in many respects a highly personal one. It differs as a problem in its total pattern and in its details as seen by the young person himself. It is personal in the nature and the degree of inner conflict and travail it involves and it is individual in the pattern of its resolution—in changes, which may ensue, in the direction of behaviour, overt, emotional, social.

Indeed, a goal toward which adolescent self-examination is implicitly or explicitly striving is the discovery and the expression of individuality. The young person, with his newly acquired ability to think critically about his own thinking—his feelings, desires, aspirations—strives to discover the various facets of himself and then to be himself through self-expression, all this in relation to the world of people to which he must relate and harmonize himself.

It is, of course, at this point of relating to society that youth encounters his greatest difficulty. He is by nature inclined to be idealistic. He wants love and trust in his relations with others. He looks for truly loving and valuing behaviour in the mature models about him. He looks for honesty, trustworthiness, genuineness, integrity, but too often and in too many places he sees quite the opposite in human interaction in business, politics, government, education and organized religion.

These discrepancies between his ideals and what he finds in society sometimes lead a youth to become critical and contentious. The erroneous assumption of a general "conflict of generations" at the time of adolescence has developed.

It is true, of course, that adolescent contentiousness sometimes takes the form of a rebellion against parental restrictions or an open conflict of ideas between the youngster and his parents. The parents, their ideas and motivational systems, sometimes become equated with those of adult society in general and the conflict may lead to the youngster's complete rejection of the adult world. He turns to his peers for support and like-minded peer groups attempt to establish subcultures of their own.

Young people with different temperamental natures and experiential backgrounds, however, react differently to essentially similar situations. Some may lead out in open protest as, *for example,* in "activist" movements among students. Others are inclined to react with hostility and destructiveness. This type of "conflict," of course, is one, which has stirred society to action. The rise of "juvenile delinquency" and crime has led not only to corrective and punitive measures against the young offenders but also to efforts to determine some of the causes of delinquency and to institute preventive measures such as providing better facilities for the constructive use of leisure time.

Nature of the "Adolescent Crisis"

Our concern in this chapter thus, far has been primarily with the developmental aspects of the period we call adolescence. We have noted at points certain essential relations between these various aspects of change the relation between physical structure and physiological functioning, *for example,* and the "psychological accompaniments" of physical and physiological change. All of these changes, taking place as an integrated process in a relatively short period of time, constitute a complex and often difficult problem-perhaps more aptly, a set of difficult problems-for the child and for those about him. There are, of course, the problems of psychological adjustment which beset the young person himself" problems of parents and others who relate to him, trying to understand more clearly his specific needs and the broader societal problems-problems of education, problems of environmentally providing for healthy adjustment and preventing maladjustments.

The degree to which such problems can be intelligently coped with obviously depends upon the clarity with which the real nature of the "adolescent crisis," with its complex dynamics and its many variations, is comprehended.

The Traditional Conception

Many studies of adolescence, psychological, sociological and anthropological, have been made through the years. Adolescence has been studied as a period of development, as a crisis period in the child's life and in his interpersonal and social relations and also as a social problem confronted by each succeeding adult generation. Consequently, an overall "traditional" conception of the nature of adolescence has become quite thoroughly established. The following statement is an attempt to characterize that traditional conception.

As a result of physical growth changes and changes in physiological functioning the child experiences a complex of psychological accompaniments—feelings, emotional stirrings, motivating "drives"—that are completely new and strange to him. He even feels strange to himself as a physical being. He is disturbed by the new quality of his emotions. He experiences an upsurge of erotic desires along with the resurgence of the old so-called Oedipus complex-his earlier erotic attachment to his mother and rivalry feelings toward his father. These feelings he finds inconsistent with the ideals and standards of right and wrong, which he had previously internalized and so there is conflict. Anxieties, feelings of guilt, apprehension, resentment arise. He finds himself in a state of bewilderment, a state of "storm and stress."

His sense of growing up also makes him realize that he is approaching adulthood and therefore he must soon assume that role and meet the expectations associated with it. He consequently becomes more concerned with the nature of the adult world and with how he will be able to relate to it and function in it. But with his idealism he is likely to see little about the world to be pleased with. He is likely to become disenchanted with the adult generation (the "generation gap"). He begins to feel pressure to detach himself from his family physically and emotionally. He must discover "a means of escaping his dependent status in the family and even more urgently the dimly recognized drives and feelings toward his parents. This is the psychological irritation which pushes the child from home, leading him to negotiate or battle with the parents for greater freedom."

All of these new psychological experiences combine to lead the youngster to a serious self-examination. He must try to make some sense of it all. He is faced with ego identity crisis—the problem of relating past to present, to future. In facing this problem he sees no source of help and support except age peers—others who presumably also are dealing with their identity problems. Hence the peer group takes on its crucial importance to him. The peer group and culture supplant the family as an anchor. It "provides a haven in which the delicate task of self-exploration and self-definition can be accomplished".

The reports of social workers and clinicians who daily work with individual problems of disturbed youngsters and their families, of course, differ in details, but in general terms they correspond rather closely to the above picture of the adolescent in crisis. Theoretical writers, generally drawing on clinical studies for their data, also describe the adolescent predicament, with certain individual and cultural variations, in the same general terms.

The general implication in the literature about adolescence is that this growing-up crisis is universal and not to be escaped—that all youth actually experience the "storm and stress" syndrome, disenchantment with the "establishment" and the need for detachment from family in their drive for autonomy. Although most writers suggest that the problem of the female adolescent may be different in detail from that of the male, the description generally is spelled out in terms of the male youth. What seems to be indicated here is the need for more "normative" data based upon samples of youth, which more fairly and adequately represent the general adolescent segment of the population. Clinical studies, without doubt, most accurately portray the nature of the problem or problems with which the clinician works, but clinicians do not work with individuals who are not embroiled in problems of maladjustment serious enough to require therapy. They deal with pathology and their descriptions and generalizations may not apply at all to the non-pathological majority. The need is for more careful studies that are not subject to the "pathologist's error."

Study of "Normal" Youth

Unfortunately, not many extensive normative studies have been made of the psychological development and adjustments of young people during adolescence. The major longitudinal projects of the past half century, of course, have made important contributions, particularly in their portrayal of individual psychological development in relation to family and other cultural influences. But these studies of relatively small and rather highly "selected" groups may not be fairly applicable to any general population. Moreover, they often furnish relatively little data, which are directly comparable from case to case concerned specifically with the problems and development of adolescence. Consequently the definitive generalizations of these studies, valuable and enlightening as they are, may not aptly portray adolescence generally.

Some rather extensive surveys of the attitudes and adjustments of adolescent boys and girls have been conducted by the Survey Research Center of the University of Michigan (Douvan and Adelson, 1966). The data for these studies were obtained by interview involving some 1045 boys and 2005 girls from school grades 6 through 12. Careful sampling techniques were used in the selection of the subjects. The individual interviews were from one to four hours in duration, conducted at the schools. This means, of course, that the study was limited to students in school, which rather definitely "under-represents 'problem' children in the population at large".

The findings of this research, then, have reference specifically to ordinary non-pathological, school-attending youth of the late 1950's. The following paragraphs present a brief summary of the results and the conclusions based upon them.

Forward Thrust of Adolescence

The tendency in ordinary American youth, according to these studies, generally was not to turn their backs upon the prospect of a life in the world of adults—to become "alienated" and without a goal in relation to society as it exists. On the contrary, the adolescent "is both pushed and pulled toward a future" by forces within him. He does experience the organically aroused "psychic conflicts" associated with puberal change. The "regressive dangers" sensed in connection with these conflicts "assemble the power of the past to urge the child to leave family and childhood". There was, in other words, the urge toward the independence of adulthood. It was, however, a pull in the positive, forward direction rather than a denial, a rejection of something regarded as inherently evil to be repudiated. To the American adolescent, then, the future was by no means a remote or irrelevant prospect.

Sex Differences

This forward thrust was found to be common to both boys and girls, but its style and focus differ markedly between the sexes. Boys were concerned with quite different kinds of future activity than were girls. Boys' choices were concrete and reality oriented. They thought in terms of preparation for specific work roles in relation to their own interests, tastes and capabilities. Their choices, as a rule, were not colored by "dreams of glory." They thought in terms of work not much above the level of their own fathers occupations. The findings also suggested a relation between the clarity of vocational goals, along with realistic notions of what it takes to achieve those goals and patterns of personal adjustment in their current lives.

Girls, on the other hand, according to this study were not strongly oriented toward jobs and vocational activities.

Their focus was upon the interpersonal and social aspects of future life. Marriage was the common goal, with special concern in their thinking with the roles of wife and mother. However, they did have ideas about "instrumental acts" appropriately leading to the primary goals. Thus, their choices of vocations were those which express their feminine interests and which were likely to place them in settings, which will bring them in contact with prospective husbands. Again, the results indicated a relation between clarity of goal aspiration and personal adjustment.

A clear concept of her adult femininity, of feminine goals and interpersonal skills, functions for the girl like the vocational concept for the boy. It bridges the worlds of adolescence and adulthood, brings the future concretely into current life and allows the future to contribute meaning and organization to adolescent activities and interests. Girls who have relatively clear notions about the goals in adult femininity show a high degree of personal integration. Those girls who specifically reject a feminine future are troubled adolescents.

The authors see the expression of these sex differences in their characteristic mobility aspirations. Those of boys typically were very realistically oriented. They were seen as "the concrete expression of a boy's faith in himself. The goal he chooses is realistic in light of his talent and opportunities, but is not overblown. And is cast in the phrasing of reality-what is the job like?" The girl's mobility aspirations were less tied to reality. Since their future mobility up or down depends upon the husband they get they "may as well dream big".

Another difference, which showed up between the sexes was in relation to the felt need to detach oneself from the family and to develop a measure of independence. In the first place, the researchers saw in general a less dramatic and less conflictful process "than tradition and theory hold." They also noted a significant sex difference. The urge to be free and "to be one's own master is almost exclusively a masculine stirring". The girls of the study, up to the age of 18, showed no significant drive for independence.

Closely associated with the urge for independence was the issue of the importance of the peer group. A conclusion of this study was that in theory the importance of the peer group to the ordinary adolescent has been much exaggerated. There was also an interesting sex difference in allegiance to the group. The boys more frequently saw "the gang" as a loyal group who could give support to its members. Girls, they concluded, may use their peer relations in quite a different way. Two-person friendships were more common. The girl more frequently needed and sought the loyalty and support of a best friend. This person-to-person shared intimacy transcended the importance of the group in meeting needs of girls.

Identity Problem

With respect to the search for personal identity Douvan and Adelson (1966), from the results of their studies, characterized this problem also largely in terms of differences they noted between boys and girls. They conclude that "there is not one adolescent crisis, but two major and clearly distinct ones—the masculine and the feminine".

The identity problem is also phased differently for boys and girls in our culture and the distinction again revolves around their different requirements for object love and for autonomy. We have noted that feminine identity forms more closely about capacity and practice in the personal arts and we have seen in our findings evidence that the girl's ego integration co-varies with her interpersonal development. Masculine identity, in contrast, focuses about the capacity to handle and master nonsocial reality, to design and win for oneself an independent area of work which fits one's individual talents and tastes and permits achievement of at least some central personal goals. The boy's ego development at. adolescence already bears the mark of this formulation and reflects his progress· in mastering it. Identity is for the boy a matter of individuating internal basis for action and defending these against domination of others. For the girl it is a process of finding and defining the internal and individual through attachments to others.

The overall view of the common adolescent crisis presented by Douvan and Adelson, as they found it among ordinary school-attending American adolescents in the late 1950s, clearly differs in significant ways from the "traditional" view. In the latter view the youngster is described as becoming intensely concerned with ethical values, religious beliefs and political ideologies. His need to detach himself from his family comes about largely because of a value crisis—a conflict between his idealism and the views of his elders, the "establishment." The interview material of these investigators supports the view that" American adolescents are, on the whole, not deeply involved in ideology, nor are they prepared to do much individual thinking on value issues of any generality" (p. 353), probably because to do so would endanger their relations in the community and complicate their personal problems of ego synthesis. These defenses, they concluded, tend to curtail experience and actually to hamper growth and self-differentiation. The traditional pattern of conflict and detachment "can be found in adolescence, but it is found in a bold, sometimes stubborn and often unhappy minority".

***

3

Cognitive Development

Piaget's 'Four levels of development' are **(1)** infancy, **(2)** preschool, **(3)** childhood and **(4)** adolescence. Each stage is characterized by a general cognitive structure that affects all of the child's thinking (a structuralist view influenced by philosopher Immanuel Kant). Each stage represents the child's understanding of reality during that period and, each but the last is an inadequate approximation' of, reality. Development from one stage to the next is thus, caused by the accumulation of errors in the child's understanding of the environment; this accumulation eventually causes such a degree of cognitive disequilibrium that thought structures require reorganizing.

The four development stages are described in Piaget's theory as:

1. *Sensorimotor Stage:* From birth to age 2. Children experience the world through movement and senses (use five senses to explore the world). During the sensorimotor stage children are extremely egocentric, meaning they cannot perceive the world from others' viewpoints. The sensorimotor stage is divided into six sub-stages: "**(1)** simple reflexes; **(2)** first habits and primary circular reactions; **(3)** secondary circular reactions; **(4)** coordination of secondary circular reactions; **(5)** tertiary Circular reactions, novelty and curiosity; and **(6)** internalization of schemes." Simple reflexes is from birth to 1 month old. At this time

infints use reflexes such as rooting and sucking. First habits and primary circular reactions is from 1 month to 4 months old. During this time infants learn to coordinate sensation and two types of scheme (habit and circular reactions). A primary circular reaction is when the infant tries to reproduce an event that happened by accident (ex: Sucking thumb). The third stage, secondary circular reactions" occurs when the infant is 4 to 8 months old. At this time they become aware of things beyond their own body; they are more object oriented. At this time they might accidentally shake a rattle and continue to do it for sake of satisfaction. Co-ordination of secondary circular reactions is from 8 months to 12 months old. During this stage they can do things intentionally. They can now combine and recombine schemes and try to reach a goal (ex: Use a stick to reach something). They also understand object permanence during this stage. That is, they understand that objects continue to exist even when they can't see them. The fifth stage occurs from 12 months old to 18 months old. During this stage infants explore new possibilities of objects; they try different things to get different results. During the last stage they are 18 to 24 months old. During this stage they shift to symbolic thinking. Some followers of Piaget's studies of infancy, such as Kenneth Kaye argue that his contribution was as an observer of countless phenomena not previously described, but that he didn't offer explanation of the processes in real time that cause those developments, beyond analogizing them to broad concepts about biological adaptation generally.

2. *Preoperational stage:* From ages 2 to 7 (magical thinking pre-dominates. Acquisition of the motor skills). Egocentrism begins strongly and then weakens. Children cannot conserve or use logical thinking.

3. *Concrete operational stage:* From ages 7 to 12 (children begin to think logically but are very concrete in their thinking). Children can now conserve and think logically but only with practical aids. They are no longer egocentric.
4. *Formal operational stage:* From age 12 onwards (development of abstract reasoning). Children develop abstract thought and can easily conserve and think logically in their mind.

The Developmental Process

Piaget provided no concise description of the development process as a whole. Broadly speaking it consisted of a cycle:

- The child performs an action which has an effect on or organizes objects and the child is able to note the characteristics of the action and its effects.
- At the same time, the child is able to identify the properties of objects by the way different kinds of action affect them. This is the process of "empirical abstraction".
- Through repeated actions, perhaps with variations or in different contexts or on different kinds of objects, the child is able to differentiate and integrate its elements and effects. This is the process of "reflecting abstraction" (described in detail in Piaget 2001).
- By repeating this process across a wide range of objects and actions, the child establishes a new level of knowledge and insight. This is the process of forming a new "cognitive stage". This dual process allows the child to construct new ways of dealing with objects and new knowledge about objects themselves.
- However, once the child has constructed these new kinds of knowledge, he or she starts to use them to create still more complex objects and to carry out

> still more complex actions. As a result, the child starts to recognize still more complex patterns and to construct still more complex objects. Thus, a new stage begins, which will only be completed when all the child's activity and experience have been re-organized on this still higher level.

This process is not wholly gradual, however. Once a new level of organization, knowledge and insight proves to be effective fit will quickly be generalized to other areas. As a result, transitions between stages tend to be rapid and radical and the bulk of the time spent in a new stage consists of refining this new cognitive level. When the knowledge that has been gained at one stage of study and experience leads rapidly and radically to a new higher stage of insight, a gestalt is said to have occurred.

It is because this process takes this dialectical form, in which each new stage is created through the further differentiation, integration and synthesis of new structures out 'of the old, that the sequence of cognitive stages are logically necessary rather than simply empirically correct. Each new stage emerges only because the child can take for granted the achievements of its predecessors and yet there are, still more sophisticated forms of knowledge and action that are capable of being developed.

Because it covers both how we gain knowledge about objects and our reflections on our own actions, Piaget's model of development explains a number of features of human knowledge that had never previously been accounted for. *For example,* by showing how children progressively enrich their understanding of things by acting on and reflecting on the effects of their own previous knowledge, they are able to organize their knowledge in increasingly complex structures. Thus, once a young child can consistently and accurately recognize different kinds of animals, he or she then. acquires the ability to organize the different kinds into higher groupings such as "birds", "fish" and so on. This is significant because they are now able to know things about a new animal simply on the basis of the fact that it is a bird - *for example,* that it will lay eggs.

At the same time, by reflecting on their own actions, the child develops an increasingly sophisticated awareness of the "rules" that govern in various ways. *For example,* it is by this route that Piaget explains this child's growing awareness of notions such as "right", "valid", "necessary", "proper" and so on. In other words, it is through the process of objectification, reflection and abstraction that the child constructs the principles on which action is not only effective or correct but also *justified.*

One of Piaget's most famous studies focused purely on the discriminative abilities of children between the ages of two and a half years old and four and a half years old. He began the study by taking children of different ages and. placing two lines of sweets, one with the sweets in a line spread further apart and one with the same number of sweets in a line placed more closely together. He found that, "Children between 2 years, 6 months old and 3 years, 2 months' old correctly discriminate the relative number of objects in two rows; between 3 years, 2 months and 4 years, 6 months they indicate a longer row with fewer objects to have "more"; after 4 years, 6 months they again discriminate correctly" *(Cognitive Capacity of Very Young Children,* p. 141). Initially younger children were not studied, because if at four years old a child could not conserve quantity, then a younger child presumably could not either. The results show however, that children that are younger than three years and two months have quantity conversation, but as they get older they lose this quality and do not recover it until four and a half years old. This attribute may be lost due to a temporary inability to solve because of an overdependence on perceptual strategies, which correlates more candy with a longer line of candy, or due to the inability for a four year old to reverse situations.

By the end of this experiment several results were found. First, younger children have a discriminative ability that shows the logical capacity for 'Cognitive operations exists earlier than acknowledged. This study also reveals that young children can be equipped with certain qualities for cognitive operations, depending on how logical the structure of the task is. Research also shows that children develop explicit. understanding at age 5 and as a result, the child will count the sweets to decide which has more. Finally the study found" that overall quantity conservation is not a basic characteristic of humans' native inheritance.

Challenges

Piaget's theory, however, vital in understanding child psychology, did not go without scrutiny. A main figure in the ratification of Piaget's ideas was the Russian psychologist Lev Vygotsky. Vygotsky stressed the importance of a child's cultural background as an effect to the stages of development. Because different cultures stress different social interactions, this challenged Piaget's theory that the hierarchy of learning development had to develop in succession. Vygotsky introduced the' term Zone of proximal development as an overall task a child would have to develop that would be too difficult to develop alone.

Curiously, Piaget had published a novel at the age of 20, before he'd begun any research in psychology, in which he stated what would later be the "conclusions" from decades of studying the development of intelligence in children.

Genetic Epistemology

According to Jean Piaget, genetic epistemology "attempts to explain knowledge and in particular scientific knowledge, on the basis of its history, its sociogenesis and especially the psychological origins of the notions and operations upon which it is based". Piaget believed he could test epistemological questions by studying the development of thought and action in children. As a result Piaget created a field known as genetic epistemology with its own methods and problems. He defined this field as the study of child development as a means of answering epistemological questions. His exploration of genetic epistemology is divided into four different stages: **(1)** the sociological model of development, **(2)** the biological model of intellectual development, **(3)** the elaboration of the logical model of intellectual development and **(4)** the study of figurative thought.

Stage 1 : The Sociological Model of Development

Piaget first developed this stage in the 1920's. He investigated the hidden side of children's minds. Piaget proposed that children moved from a position of egocentrism to socio-centrism. For this explanation he combined the use of psychological and clinical methods to create what he called a semi-clinical interview. He began the interview by asking children standardized questions and depending on how they answered, he would ask them a series of nonstandard questions. Piaget was looking for what he called "spontaneous conviction" so he often asked questions the children neither expected nor anticipated. In his studies, he noticed there was a gradual progression from intuitive to scientific and socially acceptable responses. Piaget theorized children did this because of the social interaction and the challenge to younger children's ideas by the ideas of those children who were more advanced.

Stage 2 : The Biological Model of Intellectual Development

In this stage, Piaget described intelligence as having two closely interrelated parts. The first part, which is from the first stage, was the content of children's thinking. The second part was the process of intellectual activity. He believed this process of thinking could be regarded as an extension of the biological process of adaptation. Adaptation has two pieces: assimilation and accommodation. To test his theory, Piaget observed the habits in his own children. He argued infants were engaging in an act of assimilation when they sucked on everything in their reach. He claimed infants transform all objects into an object to be sucked. The children were assimilating the objects to conform to their own mental structures. Piaget then made the assumption whenever one transforms the world to meet individual needs or conceptions; one is, in a way, assimilating it. Piaget also observed his children not only assimilating objects to fit their needs, but also modifying some of their mental structures to meet the demands of the environment. This is the second division of adoption known as accommodation. To start out, the infants only engaged in primarily reflex actions such as sucking, but not long after, they would pick up actual objects and put them in their mouths. When they do this, they modify their reflex response to accommodate the external objects into reflex actions. Because the two are often in conflict, they provide the impetus for intellectual development. The constant need to balance the two; triggers intellectual growth.

Stage 3 : The Elaboration of the Logical Model of Intellectual Development

In the model Piaget developed in stage three, he argued the idea that intelligence develops in a series of stages that are related to age and are progressive because one stage must be accomplished before the next can occur. For each stage of development the child forms a view of reality for that age period. At the next stage, the child must keep up with earlier level of mental abilities to reconstruct concepts. Piaget concluded intellectual development as an upward expanding spiral in which children must constantly reconstruct the ideas formed at earlier levels with new, higher order concepts acquired at the next level.

Stage 4 : The Study of Figurative Thought

Piaget studied areas of intelligence like perception and memory that aren't entirely logical. Logical concepts are described as being completely reversible because they can always get back to the starting point. The perceptual concepts Piaget studied could not be manipulated. To describe the figurative process,

Piaget uses pictures as examples. Pictures can't be separated because contours cannot be separated from the forms they outline. Memory is the' same way. It is never completely reversible. During this last period of work, Piaget and his colleague Inhelder also published books on perception, memory and other figurative processes such as learning during this last period.

Recently, Jonathan Tsou argued that Piaget's later epistemological works could serve as a remedy for the flaws in Thomas Kuhn's theory of scientific revolutions. However, this criticism missed some of the history between them, as well as the existence of a "lost manuscript" by Kuhn (currently held at the University of Chicago) that was to address his critics' concerns. It is noted, however, that the implications of his later' work remain largely unexamined.

The Physical Microstructure of "Schemes"

In his *"Biology and Knowledge"* 0967+ / French 1965), Piaget tentatively hinted at possible physical embodiments for his abstract "scheme" entities. At the time, there was much, talk and research about RNA as such an agent of learning and Piaget considered some of the evidence. However, he did not offer any firm conclusions and confessed that this was beyond his area of expertise.

Piaget died in 1980 and by then the RNA theory had lost its appeal. One key reason was this: *Until recently, nearly all RNA was assumed to be wholly devoted to producing protein; and such protein did not fit in with the evidence about learning.* However, in about 2000 it became clear that only about 3% of RNA was thus, employed and the remaining "non-coding" RNA (ncRNA)—the 97%—was thus, available for other tasks, including possible embodiments of Piaget's "scheme" elements. (Traill, 2005b/2008).

It has still not been established that this ncRNA scheme-basis is true. (There are methodological and other problems (Traill, 2000). However, some interesting theoretical advances have been made possible because of that theoretical development, including some unexpected explanations in various disciplines. In particular such stereotyped behavioural traits (capable of later modification or re-configuration); and Piagetian/Darwinian *trial-and-error* amongst massive populations of such' entities.

It also, implies the need for a significant amount of organized short-range *infra-red* activity and that also yields some unexpected explanations in its own right. *e.g.*, (iii) it possibly accounts for an anomaly in the capability of the optic nerve - which appears to carry much more information than it seems capable of (judged in terms of traditional mechanisms alone). See optic nerve, appendix and (iv) it seems likely to explain the century-old mystery of how *myelin geometry* is controlled. (Traill, 2005a).

Influence

Despite ceasing to be a fashionable psychologist, the magnitude of Piaget's continuing influence can be measured by the global scale and activity of the Jean Piaget Society, which holds annual conferences and attracts very large numbers of participants. His theory of cognitive development has proved influential in many different areas:

- Developmental psychology
- Philosophy
- Historical studies of thought and cognition
- Education and Morality
- Primatology
- Artificial Intelligence (AI)

Developmental Psychology

Piaget is without doubt one of the most influential developmental psychologists, influencing not only the work of Lev Vygotsky and of Lawrence Kohlberg but whole generations of eminent academics. Although subjecting his ideas to massive scrutiny led to innumerable improvements and qualifications of his original model and the emergence of a plethora of neo-Piagetian and post-Piagetian variants, Piaget's original model has proved to be remarkably robust (Lourenco and Machado 1996).

Primatology

Piaget's models of cognition have also been applied outside the human sphere and some primatologists assess the development and abilities of primates in terms of Piaget's model.

Historical Studies of Thought and Cognition

Historical changes of thought have been modeled in Piagetian terms. Broadly speaking these models have mapped changes in morality, intellectual life and cognitive levels against historical changes (typically in the complexity of social systems).

Notable examples include:

- Charles Radding's theory of the medieval intellectual development.
- Christopher Hallpike's studies of changes in cognition and moral judgment in pre-historical, archaic and classical periods ... (Hallpike 1979, 2004)
- Don Lepan's theory of the origins of modern thought and drama.
- James W. Fowler's stages of faith development.
- Jurgen Habermas's reworking of historical materialism.
- Kieran Egan's stages of understanding.
- Lawrence Kohlberg's stages of moral development.
- Michael Horace Barnes' study of the co-evolution of religious and scientific thinking.
- Peter Damerow's theory of prehistoric and archaic thought.
- Suzy Gablik's stages of art history.

Non Human Development

Neo-Piagetian stages have been applied to the maximum stage attained by various animals. *For example,* spiders attain the circular sensory motor stage, coordinating actions and perceptions. Pigeons attain the sensory motor stage, forming concepts.

Origins

The origans of human intelligence have also been studied in Piagetian terms. Wynn (1979, 1981) analysed Acheulian and Oldowan tools in terms of the insight into spatial relationships required to create each kind. On a more general level, Robinson's *Birth of Reason* (2005) suggests a large-scale model for the emergence of a Piagetian intelligence.

Education and Development of Morality

During the 1970s and 1980s, Piaget's works also inspired the transformation of European and American education, including both theory and practice, leading to a more 'childcentred' approach. In *Conversations with Jean Piaget,* he says: "Education, for most people, means trying to lead the child to resemble the typical adult of his society ... but for me and no one else, education means making creators .... You have to make inventors, innovators not conformists" (Bringuier, 1980).

Piaget's influence is strongest in early education and moral education.

His theory of cognitive development can be used as a tool in the early childhood classroom. According to Piaget, children developed best in a classroom with interaction.

Piaget believed in two basic principles relating to moral education: That children develop moral ideas in stages and that children create their conceptions of the world. According to Piaget, ('the child is someone who constructs his own moral world view, who forms ideas about right and wrong and fair and unfair, that are not the direct product of adult teaching and that are often maintained in the face of adult wishes to the contrary" (Gallagher, 1978, p. 26). Piaget believed that children made moral judgments based on their own observations of the world.

Piaget's theory of morality was radical when his book, The Moral Judgment of the Child, was published in 1932 for two reasons: his use of philosophical criteria to define morality (as universalizable, generalizable and obligatory) and his rejection of equating cultural norms with moral norms. Piaget, drawing on Kantian theory, proposed that morality developed out of peer interaction and that it was autonomous from authority mandates. Peers, not parents, were a key source of moral concepts such as equality, reciprocity and justice.

Piaget attributed different types of psychosocial processes to different forms of social relationships, introducing a fundamental distinction between different types of said relationships... Where there is constraint because one participant holds more power than the other the relationship is asymmetrical and, importantly, the knowledge that can be acquired by the dominated participant takes on a fixed and inflexible form. Piaget refers to this process as one of social transmission, illustrating it through reference to the way in which the elders of a tribe initiate younger members into the patterns of beliefs and practices of the group. Similarly, where adults exercise a dominating influence over the growing child, it is through social transmission that children can' acquire knowledge. By contrast, in cooperative relations, power is more evenly distributed between participants so that a more symmetrical relationship emerges. Under these conditions, authentic forms of intellectual exchange become possible; each partner has the freedom to project his or her own thoughts, consider the positions of others and defend his or her own point of view. In such circumstances, where children's thinking is not limited by a dominant influence, Piaget believed "the reconstruction of knowledge", or favorable conditions for the emergence of constructive solutions to problems, exists. Here, the knowledge that emerges is open, flexible and regulated by the logic of argument rather than being determined by an external. authority. In short, cooperative relations provide the arena for the emergence of operations, which for Piaget requires the absence of any constraining influence and is most often illustrated by the relations that form between peers (for more on the importance of this distinction see Duveen & Psaltis, in press; Psaltis & Duveen, 2006, 2007).

Philosophy

Some have taken account of Piaget's work. *For example,* the philosopher and social theorist Jurgen Habermas has incorporated Piaget into his work, most notably in *The Theory of Communicative Action.* The philosopher Thomas Kuhn credited Piaget's work in helping him understand the transition between modes of thought which characterized his theory of paradigm shifts. Shortly before his death (September, 1980), Piaget was involved in a debate about the relationships between innate and acquired features of language, at the Centre Royaumont pour une Science de I'Homme, where.he discussed his point of view with the linguist Noam Chomsky as well as Hilary Putnam and Stephen Toulmin.

Artificial Intelligence

Piaget also had a considerable effect in the field of computer science and artificial intelligence. Seymour Papert used Piaget's work while developing the Logo programming language. Alan Kay used Piaget's theories as the basis for the Dynabook programming system concept, which was first discussed within the confines of the Xerox Palo Alto Research Center, or Xerox PARC. These discussions led to the development of the Alto prototype, which explored for the first time all the elements of the Graphical User Interface (GUI) and influenced the creation of user interfaces in the 1980s and beyond.

❋❋❋

4

Social and Emotional Development

Developmental psychologists who are interested in social development examine how individuals develop social and emotional competencies. *For example,* they study how children form friendships, how they understand and deal with emotions and how identity develops. Research in this area may involve study of the relationship between cognition or cognitive development and social behaviour.

Research Design

Developmental psychologists have a number of methods to study changes in individuals over time.

In a longitudinal study, a researcher observes many individuals born at or around the same time (a cohort) and carries out new observations as members of the cohort age. This method can be used to draw conclusions about which types of development are universal (or normative) and occur in most members of a cohort. Researchers may also observe ways in which development varies between individuals and hypothesize about the causes of variation observed in their data. Longitudinal studies often require large amounts of time and funding, making them unfeasible in some situations. Also, because members of a cohort all experience historical events unique to their generation, apparently normative developmental trends may in fact be universal only to their cohort.

In a cross-sectional study, a researcher observes differences between individuals of different ages at the same time. This generally requires less resources than the longitudinal method and because the individuals come from different cohorts, shared historical events are not so much of a confounding factor. By the same token, however, cross-sectional research may not be the most effective way to study differences between participants, as these differences may result not from their different ages but from their exposure to different historical events.

An accelerated longitudinal design or cross-sequential study or cohort-sequential design combines both methodologies. Here, a researcher observes members of different birth cohorts at the same time and then tracks all participants over time, charting changes in the groups. By comparing differences and similarities in development, one can more easily determine what changes can be attributed to individual or historical environment and which are truly universal. Clearly such a study can be even more resource-consuming than a longitudinal study.

Additionally, these are all correlational, not experimental, designs and so one cannot readily infer causation from the data they yield. Nonetheless, correlational research methods are common in the study of development, in part due to ethical concerns. In a study of the effects of poverty on development, for instance, one cannot easily randomly assign certain families to a poverty condition and others to an affluent one and so observation alone has to suffice.

Stages of Development

The prenatal development of human beings is viewed in three separate stages, which are not the same as the trimesters of a woman's pregnancy:

1. Germinal (conception through week 2).
2. Embryonic (weeks 3 through 8).
3. Fetal (week 9 through birth).

The germinal stage begins when a sperm penetrates an egg in the act of conception (normally the result of sexual inter-course between a man and a woman). At this point a zygote is formed. Through the process of mitosis, the cells divide and double.

The embryonic stage occurs once the zygote has firmly implanted itself in the uterine wall. It is in this stage that the vital organs are formed and while the external body is still extremely dissimilar from an adult human, some features such as eyes and arms and eventually ears and feet, become recognizable.

The fetal period is the pre-natal period when the brain has its greatest development, becoming more and more complex over the last few months.

During pregnancy there is a risk to the developing child from drugs and other teratogens, spousal abuse and other stress on the mother, nutrition and the age of the mother. Genetic testing prior to pregnancy is also increasingly available. Three methods of determining fetal defects and health include the ultrasound, amniocentesis and chorionic villus sampling. Although difficult, some methods of treating fetal disorders have been developed, both surgical and drug based.

Infancy

From birth until the onset of speech, the child is referred to as an infant. Developmental psychologists vary widely in their assessment of infant psychology and the influence the outside world has upon it, but certain aspects are relatively clear.

While no agreement has yet been reached regarding the level of stimulation an infant requires, a normal level of stimulation is very important and a lack of stimulation and affection can result in learning difficulties and a host of other developmental and social disorders Some feel that classical music, particularly Mozart is good for an infant's mind. While some tentative research has shown it to be helpful to older children, no conclusive evidence is available involving infants.

The majority of a newborn infant's time is spent in sleep. At first this sleep is evenly spread throughout the day and night, but after a couple of months, infants generally become diurnal.

Infants can be seen to have 6 states, grouped into pairs:

- Quiet sleep and active sleep (dreaming, when REM occurs).
- Quiet waking and active waking.
- Fussing and crying.

Infants respond to stimuli differently in these different states. Habituation is frequently used in testing psychological phenomenon. Both infants and adults attend less as a result of consistent exposure to a particular stimulus. The amount of time spent attending to an alternate stimulus (after habituation to the initial stimulus) is indicative of the strength of the remembered percept of the previous stimulus, or dishabituation.

Habituation is used to discover the resolution of perceptual systems, *for example,* by habituating a subject to one stimulus and then observing responses to similar ones, one can detect the smallest degree of difference that is detectable by the subject.

Infants have a wide variety of reflexes, some of which are permanent (blinking, gagging) and others transient in nature. Some have obvious purposes, some are dearly vestigial and some do not have obvious purposes. Primitive reflexes reappear in adults under certain conditions, such as neurological conditions like dementia or traumatic lesions. A partial list of infantile reflexes includes:

1. **Moro reflex or startle reflex:**
    - Startle.
    - Spreading out the arms (adduction).
    - Unspreading the arms (abduction).
    - Crying (usually).

2. Tonic neck reflex or fencer's reflex.
3. **Stepping reflex, step-up reflex:** Can be initiated if you support the infant upright from its armpits below a given surface so the baby lifts its foot and steps up on the surface (like climbing a stair).
4. **Rooting reflex, sucking reflex, suckling reflex:** Can be initiated by scratching the infant's cheek; the reaction is pursing of the lips for sucking.
5. **Parachute reflex:** The infant is suspended by the trunk and suddenly lowered as if falling for an instant. The child spontaneously throws out the arms as a protective mechanism. The parachute reflex appears before the onset of walking.
6. **Grasp reflex:** Can be initiated by scratching the infant's palm.
7. **Plantar reflex or Babinski reflex:** A finger is stroked firmly down the outer edge of the baby's sole; the toes spread and extend out.

Infants have significantly worse vision than older children. Infant sight, blurry in early stages, improves over time. Infants less than 2 months old are thought to be colour blind.

Hearing is well-developed prior to birth, however and a preference for the mother's heartbeat is well established. Infants are fairly good at detecting the direction from which a sound comes and by 18 months their hearing ability is approximately equal to that of adults.

Smell and taste are present, with infants having been shown to prefer the smell and taste of a banana, while rejecting the taste of shrimp. There is good evidence for infants preferring the smell of their mother to that of others.

Infants have a fully developed sense of touch at birth and the myth believed by some doctors even today that infants feel no pain is inaccurate. Doctors are slowly becoming aware of the need for pain prevention for newborns.

Piaget asserted that there were several sensorimotor stages within his broader Theory of cognitive development.

- The first sub-stage occurs from birth to six weeks and is associated primarily with the development of reflexes. Three primary reflexes are described by Piaget: sucking of objects in the mouth, following moving or interesting objects with the eyes and closing of the hand when an object makes contact with the palm (palmar grasp). Over these first six weeks of life, these reflexes begin to become voluntary actions; *for example,* the palmar reflex becomes intentional grasping.
- The second sub-stage occurs from four to nine months and is associated primarily with the development of co-ordination between vision and prehension. Three new abilities occur at this stage: intentional grasping for a desired object, secondary circular reactions and differentiations between ends and means. At this stage, infants will intentionally grasp the air in the direction of a desired object, often to the amusement of friends and family. Secondary circular reactions, or the repetition of an action involving an external object occur begin; *for example,* moving a switch to turn on a light repeatedly. The differentiation between means also occurs. This is perhaps one of the most important stages of a child's growth as it signifies the dawn of logic. Towards the late part of this sub-stage infants begin to have a sense of object permanence, passing the A-not-B error test.
- The third sub-stage occurs from six weeks to four months and is associated primarily with the development of habits. Primary circular reactions or

repeating of an action involving only ones own body begin. An example of this type of reaction would involve something like an infant repeating the motion of passing their hand before their face. Also at this phase, passive reactions, caused by classical or operant conditioning, can begin.

- The fourth sub-stage occurs from nine to twelve months and is associated primarily with the development of logic and the coordination between means and ends. This is an extremely important stage of development, holding what Piaget calls the "first proper intelligence." Also, this stage marks the beginning of goal orientation, the deliberate planning of steps to meet an objective.
- The fifth sub-stage is associated primarily with the beginnings of insight, or true creativity. This marks the passage into the pre-operational stage.
- The sixth sub-stage occurs from twelve to eighteen months and is associated primarily with the discovery of new means to meet goals. Piaget describes the child at this juncture as the "young scientist," conducting pseudo-experiments to discover new methods of meeting challenges.

When studying infants, the habituation methodology is an example of a method often used to assess their performance. This method allows researchers to obtain information about w hat types of stimuli an infant is able to discriminate. In this paradigm, infants are habituated to a particular stimulus and are then tested using different stimuli to evaluate discrimination. The critical measure in habituation is the infants' level of interest. Typically, infants prefer stimuli that are novel relative to those they have encountered previously. Several methods are used to measure infants' preference. These include the high-amplitude sucking procedure, in which infants suck on a pacifier more or less depending on their level of interest, the conditioned foot-kick procedure, in which infants move their legs to indicate preference and the head-turn preference procedure, in which the infant's level of interest is measured by the amount of time spent looking in a particular direction. A key feature of all these methods is that, in each situation, the infant controls the stimuli being presented. This gives researchers a means of measuring discrimination. If an infant is able to discriminate between the habituated stimulus and a novel stimulus, they will show a preference for the novel stimulus. If, however, the infant cannot discriminate between the two stimuli, they will not show a preference for one over the other.

Object permanence is an important stage of cognitive development for infants. Numerous tests regarding it have been done, usually involving a toy and a crude barrier which is placed in front of the toy and then removed, repeatedly. In sensorimotor stages 1 and 2, the infant is completely unable to comprehend object permanence. Jean Piaget conducted experiments with infants which led him to conclude that this awareness was typically achieved at eight to nine months of age. Infants before this age are too young to understand object permanence, which explains why infants at this age do not cry when their mothers are gone. "Out of sight, out of mind." A lack of Object Permanence can lead to A-not-B errors, where children reach for a thing at a place where it should not be.

Toddler

Intelligence is demonstrated through the use of symbols, language use matures and memory and imagination are developed. Thinking is done in a non-logical, nonreversible manner. Egocentric thinking predominates.

Socially, toddlers are little people attempting to become independent at this stage, which they are commonly called the "terrible twos". They walk, talk, use the toilet and get food for themselves. Self-control begins to develop. If taking the initiative to explore, experiment, risk mistakes in trying new things and test their limits is encouraged by the caretaker(s) the child will become autonomous, self-reliant and confident. If the caretaker is over-protective or disapproving of independent actions, the toddler may begin to doubt their abilities and feel ashamed for the desire for independence. The child's autonomic development will be inhibited and be less prepared to successfully deal with the world in the future.

Early Childhood

When children attend preschool, they broaden their social horizons and become more engaged with those around them. Impulses are channeled into fantasies, which leaves the task of the caretaker to balance eagerness for pursuing adventure, creativity and self-expression with the development of responsibility. If caretakers are properly encouraging and consistently disciplinary, children are more likely to develop positive self-esteem while becoming more responsible and will follow through on assigned activities. If not allowed to decide which activities to perform, children may begin to feel guilt upon contemplating taking initiative. This negative association with independence will lead them to let others make decisions in place of them.

Childhood

In middle childhood, intelligence is demonstrated through logical and systematic manipulation of symbols related to concrete objects. Operational" thinking develops, which means actions are reversible and egocentric thought diminishes.

Children go through the transition from the world at home to that of school and peers. Children learn to make things, use tools and acquire the skills to be a worker and a potential provider. Children can now receive feedback from outsiders about their accomplishments. If children can discover pleasure in intellectual stimulation, being productive, seeking success, they will develop a sense of competence. If they are not successful or cannot discover pleasure in the process, they may develop a sense of inferiority and feelings of inadequacy that may haunt them throughout life. This is when children think of them selves as industrious or as inferior.

Adolescence

Adolescence is the period of life between the onset of puberty and the full commitment to an adult social role, such as worker, parent and/or citizen. It is the period known for the formation of personal and social identity and the discovery of moral purpose. Intelligence is demonstrated through the logical use of symbols related to abstract concepts and formal reasoning. A return to egocentric thought often occurs early in the period. Only 35% develop the capacity to reason formally during adolescence or adulthood.

The adolescent subconsciously explores questions such as "Who am I? Who do I want to be?" Like toddlers, adolescents must explore, test limits, become autonomous and commit to an identity, or sense of self. Different roles, behaviours and ideologies must be tried out to select an identity. Role confusion and inability to choose vocation can result from a failure to achieve a sense of identity.

Early Adulthood

The person must learn how to form intimate relationships, both in friendship and love. The development of this skill relies on the resolution of other stages. It may be hard to establish intimacy if one has not developed trust or a sense of identity. If this skill is not learned the alternative is alienation, isolation, a fear of commitment and the inability to depend on others.

A related framework for studying this part of the life span is that of Emerging adulthood, introduced in 2000 by Jeffrey Arnett. Scholars of emerging adulthood are interested not only in relationship development (focusing on the role of dating in helping individuals settle on a long-term spouse/

partner), but also the development of sociopolitical views and occupational choice.

Middle Age

Middle adulthood generally refers to the period between ages 40 to 65. During this period, the middle-aged experience a conflict between generativity and stagnation. They may either feel a sense of contributing to the next generation and their community or a sense of purposelessness.

Physically, the middle-aged experience a decline in muscular strength, reaction time, sensory keenness and cardiac output. Also, women experience menopause and a sharp drop in the hormone estrogen. Men do have an equivalent to menopause, it is called "Andropause," which is a hormone fluctuation with physical and psychological effects similar to menopause. Lowered testosterone levels result in mood swings and a decline in sperm count and speed of ejaculation and erection. Most men and women remain capable of sexual satisfaction after middle age.

Old Age

This stage generally refers to those over 75 years. During old age, people experience a conflict between integrity vs. despair. When reflecting on their life, they either feel a sense of accomplishment or failure.

Physically, older people experience a decline in muscular strength, reaction time, stamina, hearing, distance perception and the sense of smell. They also are more susceptible to severe diseases such as cancer and pneumonia due to a weakened immune system. Mental disintegration may also occur, leading to Dementia or Alzheimer's Disease. However, partially due to a lifetime's accumulation of antibodies, the elderly are less likely to suffer from common diseases such as the cold.

Whether or not intellectual powers increase or decrease with age remains controversial. Longitudinal studies have suggested that intellect declines, while cross-sectional studies suggest that intellect is stable. It is generally believed that crystallized intelligence increases up to old age, while fluid intelligence decreases with age.

OTHER FINDINGS

Parenting

In Western developed societies, mothers (and women generally) were emphasized to the exclusion of other caregivers, particularly as the traditional role of the father was more the breadwinner and less the direct caregiver of an infant, he has been traditionally viewed as impacting an infant indirectly through interactions with the mother of the child.

The emphasis of study has shifted to the primary caregiver (regardless of gender or biological relation), as well as all persons directly or indirectly influencing the child (the family system). The roles of the mother and father are more significant than first thought as we moved into the concept of primary caregiver.

Affirming a role for fathers, studies have shown that children as young as 15 months benefit significantly from substantial engagement with their father. In particular, a Cognitive Development 183 study in the U.S.A. and New Zealand found the presence of the natural father was the most significant factor in reducing rates of early sexual activity and rates of teenage pregnancy in girls. Covariate factors used included early conduct problems, maternal age at first childbirth, race, maternal education, father's occupational status, family living standards, family life stress, early mother-child interaction, measures of psychosocial adjustment and educational achievement, school qualifications, mood disorder, anxiety disorder, suicide attempts, violent offending and conduct disorder. Further research has found fathers have an impact on child academic performance, including involved nonresident fathers. However, father absence is associated with a range of negative outcomes for children, including child and later criminal behaviour.

Historical Antecedents

The modern form of developmental psychology has its roots in the rich psychological tradition represented by Aristotle, Tabari, Rhazes, Alhazen and Descartes. William Shakespeare had his melancholy character Jacques (in As You Like It) articulate the seven ages of man: these included three stages of childhood and four of adulthood. In the mideighteenth century Jean Jacques Rousseau described three stages of childhood: *infans* (infancy), *puer* (childhood) and *adolescence* in *Emile*: Or, On Education. Rousseau's ideas were taken up strongly by educators at the time.

In the late nineteenth century, psychologists familiar with the evolutionary theory of Darwin began seeking an evolutionary description of psychological development; prominent here was G. Stanley Hall, who attempted to correlate ages of childhood with previous ages of mankind.

A more scientific approach was initiated by James Mark Baldwin, who wrote essays on topics that included Imitation: A Chapter in the Natural History of Consciousness and Mental Development in the Child and the Race: Methods and Processes. In 1905, Sigmund Freud articulated five psychosexual stages. Later, Rudolf Steiner articulated stages of psychological development throughout human life. By the early to midtwentieth century, the work of Vygotsky and Piaget, mentioned above, had established a strong empirical tradition in the field.

ID, EGO AND SUPER-EGO

Id, ego and super-ego are the three parts of the psychic apparatus defined in Sigmund Freud's structural model of the psyche; they are the three theoretical constructs in terms of whose activity and interaction mental life is described.

According to this model, the uncoordinated instinctual trends am the "id"; the organized realistic part of the psyche is the "ego," and the critical and moralising function the "superego."

Even though the model is "structural" and makes reference to an "apparatus", the id, ego and super-ego are functions of the mind rather than parts of the brain and do not necessarily correspond one-to-one with actual somatic structures of the kind dealt with by neuroscience.

The concepts themselves arose at a late stage in the development of Freud's thought: The structural model was first discussed in his 1920 essay "Beyond the Pleasure Principle" and was formalised and elaborated upon three years later in his "The Ego and the Id." Freud's proposal was influenced by the ambiguity of the term "unconscious" and its many conflicting uses.

The terms "id," "ego," and "super-ego" are not Freud's own but are latinisations originating from his translator James Strachey. Freud himself wrote of "das Es," "das Ich," and "das Uber-Ich" -respectively, "the It," "the I," and the "Over-I" (or "Upper-I"); thus, to the German reader, Freud's original terms are more or less self-explanatory. The term "das Es" was borrowed from Georg Groddeck, a German physician to whose unconventional ideas Freud was much attracted. (Groddeck's translators render the term in English as 'the It').

Id

The Id comprises the unorganized part of the personality structure that contains the basic drives. The id acts as according to the 'pleasure principle', seeking to avoid pain or unpleasure aroused by increases in instinctual tension.

The Id is unconscious by definition. In Freud's formulation:

It is the dark, inaccessible part of our personality, what

> little we know of it we have learnt from our study of the dream-work and of the construction of neurotic symptoms and most of this is of a negative character and can be described only as a contrast to the ego. We all approach the id with analogies: We call it a chaos, a cauldron full of seething excitations... It is filled with energy reaching it from the instincts, but it has no organization, produces no collective will, but only a striving to bring about the satisfaction of the instinctual needs subject to the observance of the pleasure principle. [Freud, New Introductory Lectures on Psychoanalysis (1933)]

Developmentally, the Id is anterior to the ego; *i.e.,* the psychic apparatus begins, at birth, as an undifferentiated id, part of which then develops into a structured ego. Thus, the id:

> Contains everything that is inherited, that is present at birth, is laid down in the constitution—above all, therefore, the instincts, which originate from the somatic organization and which find a first psychical expression here (in the id) in forms unknown to us".

The mind of a newborn child is regarded as completely "id-ridden", in the sense that it is a mass of instinctive drives and impulses and demands immediate satisfaction. This view equates a newborn child with an id-ridden individual—often humorously—with this analogy: An alimentary tract with no sense of responsibility at either end.

The id is responsible for our basic drives such as food, water, sex and basic impulses. It is amoral and egocentric, ruled by the pleasure—pain principle; it is without a sense of time, completely illogical, primarily sexual, infantile in its emotional development and will not take "no" for an answer. It is regarded as the reservoir of the libido or "instinctive drive to create".

Freud divided the id's drives and instincts into two categories: life and death instincts—the latter not so usually regarded because Freud thought of it later in his lifetime. Life instincts (Eros) are those that are crucial to pleasurable survival, such as eating and copulation. Death instincts, (Thanatos) as stated by Freud, is our unconscious wish to die, as death puts an end to the everyday struggles for happiness and survival. Freud noticed the death instinct in our desire for peace and attempts to escape reality through fiction, media and substances such as alcohol and drugs. It also indirectly represents itself through aggression.

Ego

The Ego acts according to the reality principle; *i.e.*, it seeks to please the id's drive in realistic ways that will benefit in the long term rather than bringing grief.

The Ego comprises that organized part of the personality structure which includes defensive, perceptual, intellectual-cognitive and executive functions. Conscious awareness resides in the ego, although not all of the operations of the ego are conscious. The ego separates what is real. It helps us to organize our thoughts and make sense of them and the world around us.

According to Freud,

> ... The ego is that part of the id which has been modified by the direct influence of the external world ... The ego represents what may be called reason and common sense, in contrast to the id, which contains the passions ... in its relation to the id it is like a man on horseback, who has to hold in check the superior strength of the horse; with this difference, that the rider tries to do so with his own strength, while the ego uses borrowed forces [Freud, The Ego rmd the Id (1923)]

In Freud's theory, the ego mediates among the id, the super-ego and the external world. Its task is to find a balance between primitive drives and reality (the Ego devoid of morality at this level) while satisfying the id and super-ego. Its main concern is with the individual's safety and allows some of the id's desires to be expressed, but only when consequences of these actions are marginal. Ego defense mechanisms are often used by the ego when id behaviour conflicts with reality and either society's morals, norms and taboos or the individual's expectations as a result of the internalisation of these morals, norms and their taboos.

The word ego is taken directly from Latin, where it is the nominative of the first person singular personal pronoun and is translated as "I myself" to express emphasis. The Latin term ego is used in English to translate Freud's German term Das Ich, which literally means "the I".

Ego development is known as the development of multiple processes, cognitive function, defenses and interpersonal skills or to early adolescence when ego processes are emerged.

In modern-day society, ego has many meanings. It could mean one's self-esteem; an inflated, sense of self-worth; or in philosophical terms, one's self. However, according to Freud, the ego is the part of the mind which contains the consciousness. Originally, Freud had associated the word ego to meaning a sense of self; however, he later revised it to mean a set of psychic functions such as judgment, tolerance, reality-testing, control, planning, defense, synthesis of information, intellectual functioning and memory.

In a diagram of the Structural and Topographical Models of Mind, the ego is depicted to be half in the consciousness, while a quarter is in the preconscious and the other quarter lies in the unconscious.

The ego is the mediator between the id and the superego, trying to ensure that the needs of both the id and the super-ego are satisfied. It is said to operate on a reality principle, meaning it deals with the id and the super-ego; allowing them to express their desires, drives and morals in realistic and socially appropriate ways. It is said that the ego stands for reason and caution, developing with age. Sigmund Freud had used an analogy which likened the ego to a rider and a horse; the ego being the rider while the id being the horse. The horse provides the energy and the means of obtaining the energy and information needed, while the rider ultimately controls the direction it wants to go. However, due to unfavourable conditions, sometimes the horse makes its own decisions over the rocky terrain.

When the ego is personified, it is like a slave to three harsh masters: the id, the super-ego and the external world. It has to do its best to suit all three, thus, is constantly feeling hemmed by the danger of causing discontent on two other sides. It is said, however, that the ego seems to be more loyal to the id, preferring to gloss over the finer details of reality to minimise conflicts while pretending to have a regard for reality. But the super-ego is constantly watching everyone of the ego's moves and punishes it with feelings of guilt, anxiety and inferiority. To overcome this the ego employs defense mechanisms. The defense mechanisms are not done so directly or consciously. They lessen the tension by covering up our impulses that are threatening.

Denial, displacement, intellectualisation, fantasy, compensation, projection, rationalisation, reaction formation, regression, repression and sublimation were the defense mechanisms Freud identified. However, his daughter Anna Freud clarified and identified the concepts of undoing, suppression, dissociation, idealisation, identification, introjection, inversion, somatisation, splitting and substitution.

Super-Ego

The Super-ego aims for perfection. It comprises that organized part of the personality structure, mainly but not entirely unconscious, that includes the individual's ego ideals, spiritual goals and the psychic agency (commonly called 'conscience') that criticises and prohibits his or her drives, fantasies, feelings and actions.

The Super-ego can be thought of as a type of conscience that punishes misbehaviour with feelings of guilt. *For example*: having extra-marital affairs.

The Super-ego works in contradiction to the id. The Super-ego strives to act in a socially appropriate manner, whereas the id just wants instant self-gratification. The Superego controls our sense of right and wrong and guilt. It helps us fit into society by getting us to act in socially acceptable ways.

The Super-ego's demands oppose the id's, so the ego has a hard time in reconciling the two. Freud's theory implies that the super-ego is a symbolic internalisation of the father figure and cultural regulations. The super-ego tends to stand in opposition to the desires of the id because of their conflicting objectives and its aggressiveness towards the ego. The superego acts as the conscience, maintaining our sense of morality and proscription from taboos. The superego and the ego is the product of two key factors: The state of helplessness of the child and the Oedipus complex. Its formation takes place during the dissolution of the Oedipus complex and is formed by an identification with and internalisation of the father figure after the little boy cannot successfully hold the mother as a love-object out of fear of castration.

The super-ego retains the character of the father, while the more powerful the Oedipus complex was and the more rapidly it succumbed to repression (under the influence of authority, religious teaching, schooling and reading), the stricter will be the domination of the super-ego over the ego later on—in the form of conscience or perhaps of an unconscious sense of guilt (The Ego and the ld, 1923).

In Sigmund Freud's work Civilization and Its Discontents (1930) he also discusses the concept of a "cultural super-ego". The concept of super-ego and the Oedipus complex is subject to criticism for its perceived sexism. Women, who are considered to be already castrated, do not identify with the father and therefore form a weak super-ego, leaving them susceptible to immorality and sexual identity complications.

ASPECTS OF EMOTIONALITY

Emotionality has been defined in various ways and has been seen from different theoretical points of view by students of human behaviour. One of these approaches, objective in nature, is to define an emotion in terms of the external situation—the perceptual field that gives rise to it. Thus, those emotional reactions, which result from situations that are judged objectively to be dangerous would be labeled fear or terror. Likewise, reactions to confining or frustrating situations would be anger or rage and so on. This was the approach used by early investigators of emotions in infants and children. Watson (1924), *for example,* in his efforts to account for all infant emotions, defined the three emotions fear, anger and love in terms of the relatively few stimuli to which he found infants responsive. The infants response to a sudden loud noise or a sudden loss of support he labeled fear, their reactions to restraint he labeled anger and responses to being rocked, to stroking, fondling and the like, were love.

A second objective approach to the study of emotions, closely related to the first, is in terms of the observable reactions, or measurable bodily changes, that constitute emotional behaviour. Emotional behaviour from this viewpoint is both external and internal. External behaviour, of course, is quite readily observed. When the behaviour is destructive or assaultive in nature it is labeled anger or rage. When it appears as less violent it is affection or love or lust or even elation. When behaviour appears as retreat or flight it is fear.

Quite obviously, the mere observation of external behaviour without more information, including a knowledge of the eliciting stimuli, is not a sufficient basis for an adequate classification of emotion.

The internal responses involved in emotion are very complex and, of course, cannot be observed directly. Their effects, however can be measured by the use of instruments. The study of the internal, or visceral, changes involved in emotion has contributed much to our understanding. Complicated sequences and patterns of change involving the vascular system, the endocrine system and the involuntary muscular system are extremely important aspects of the complex we call emotion. These physiological reactions are controlled and regulated by the autonomic division of the nervous system. In a very general way the "approach" and "avoidance" aspects of behaviour are mediated by the parasympathetic and the sympathetic segments of the autonomic division, respectively. The internal aspects of emotional behaviour, however, are always widespread and diffuse and so far only very rough correlations with specific nervous structures have been established.

A third approach to the problem of the emotions is to regard emotion as a conscious experience. In spite of efforts on the part of some students of human behaviour to avoid a consideration of the conscious aspect of emotions, no "normal" human being—including the objective psychologist himself—can doubt the reality of the consciously experienced emotion. When we act angry in response to an anger-provoking stimulus we also feel angry.

A Response of the Total Person

In view of these three approaches to the study of emotional behaviour, an important fact stands out: Emotionality is an extremely complex area of human functioning, which involves the whole organism. The stimulus aspect cannot be neglected. Cognition is always involved. There is no emotion without either the actual perception of the provoking situation or the anticipation of it in terms of images. Normally, a situation does not provoke genuine fear or anger or jealousy or joy or elation unless it is first perceived or thought of as something about which one must be fearful or angry or jealous or joyful or elated. Even in the case of irrational anxiety or worry, image's or other symbols representing past cognitive experiences or imagined ones are being centrally manipulated.

The external behavioural aspects, the so-called expressive movements of emotion, are equally essential to the total complex. In full-fledged anger, *for example,* although actual destructive aggression may be held in check, there are, nevertheless, implicit if not overt patterns of muscular contractions and gross bodily movements that express hostility and tendency to aggression. The kinesthetic stimuli produced by these expressive behaviours give rise to sensory experiences that are important ingredients in the total experiential aspect of the complex.

Simultaneously with overt behaviour come the internal responses, which are mediated by the autonomic division of the nervous system in response to cognitive awareness of the situation. The whole pattern of changes is "designed" to prepare the organism physically for biologically appropriate activity. In the case of fear or anger, the joint action of the autonomic nervous system and the endocrine glandular system supplies the blood with additional energy-giving blood sugar and produces changes in the pattern of blood-vessel tension so as to shunt that energy-laden blood into the skeletal muscles for greater strength in meeting the "emergency." Internal organs, like skeletal muscles, are equipped with sense receptors that are stimulated by the internal activity, thus, giving rise to patterns of sensations, which add to the "unanalyzed mass" that constitutes the emotional experience.

A realization of the importance of these sensory

components arising from response-produced stimuli was what led William James, in 1890, to make his famous paradoxical statement that "we do not tremble because we are afraid but we are afraid because we tremble." According to his theory, which was also stated independently by the Danish physiologist Carl G. Lange, an emotion is the way the body feels when in a disturbed internal state and while making various overt expressive movements. The massive, unanalyzed experience arising from the bodily processes and the expressive movements is the emotion, according to James.

The important point here is that an adequate consideration of an area of human behaviour as complex as the emotions must deal with overt behavioural, as well as psychological, aspects. We would agree essentially Ausubel (1958):

Emotion may be defined as a heightened state of subjective experience accompanied by skeletal-motor and autonomic-humoral responses and by a selectively generalized state of lowered response thresholds....

The following sequential steps are involved in the instigation of an emotional response: *(a)* interpretive phase—perception or anticipation of an event that is interpreted as threatening or enhancing an ego-involved, goal, value, or attribute of self; *(b)* preparatory reactive phase consisting of a selectively generalized lowering of the particular response thresholds implicated in a given emotion; *(c)* consumatory reactive phase with subjective, autonomic-humoral and skeletal-motor components; and *(d)* a reflective reactive phase involving subjective awareness of the drive state and of visceral and skeletal responses.

Positive and Negative Emotions

As is obvious and common in everyone's experience, there are two general directional types, or categories, of emotional behaviour and experience. These types are so different qualitatively that they seem, subjectively and appear objectively, to have nothing in common. "Pleasant," "positive," and "approach" apply to the one directional type, while "unpleasant," "negative," "escape," "injure," and "destroy" apply to the other. Not only are they logical and subjective opposites but the two types of experiences also involve mutually antagonistic autonomic functioning, thus, mediating overall patterns of visceral change that are quite different. Yet the common term "emotion" applies equally well and without conflict to both categories of experience. Both types are heightened states of subjective experience. Both are extreme deviations from the comfortable, normal internal state of affairs, both being, in a sense, "emergency" reactions.

As we shall see in other connections later, the relative dominance of one or the other of these two emotional types has important implications in the child's personal development. The prevailing nature of the young child's environment will have much to, do with determining whether the social aspects of his life become attached to the positive or the negative categories of emotion.

Emotional Development

Emotional experiences and outward emotional behaviour are matters of universal, firsthand acquaintance. Daily we all experience a variety of emotions in ourselves and we are constantly meeting the problem of dealing with them in others. Underdeveloped emotional capacity, uncontrolled emotional behaviour and pathological expressions of emotion are major areas of concern among those who must deal with human problems at any level, whether individual, group, national, or international; hence, the need for greater understanding of the nature and development of emotionality.

Controlled research in this area is extremely difficult. Behaviour under emotion-arousing conditions in natural settings can be observed directly and descriptions of that behaviour can be recorded. But affectivity itself—what the individual is actually experiencing under those conditions—obviously cannot be observed objectively. Affectivity must be inferred in terms of the situation and the observed external behaviour associated with the situation. We include in the behaviour that can be observed the individual's subjective report of his feelings.

Our greatest insights into the nature of emotionality have come from the clinical approach. In the clinical setting, real emotional problems of individuals with widely varying backgrounds and experiences are studied. The clinician must, in each case, understand and deal with a total, unique emotional situation. Clinical studies have brought to focus the facts of individuality in human behaviour and the great diversity of environmental influences that affect individual emotional development. They clearly portray, *for example,* the fact that each child has his own inner strivings and frustrations, his own "self image," his attitude toward himself in relation to others. The evidence suggests that nothing gives the child more help and support in resolving his personal problems and thus, facilitating his emotional growth, than the feeling that he is really understood, that someone important to him really knows how he feels.

Congenital Affectivity

The first breath of the newborn infant appears to be an outward expression of an emotional disturbance—a "protest" against intolerable conditions. The newborn apparently keeps on breathing because breathing promotes a satisfying state of affairs. It eliminates discomfort.

During the neonatal period the baby normally shows signs of emotional excitement only when conditions, external and internal, are such as to cause him bodily discomfort or sensuous pleasure. When he feels pain from some physical disorder or external condition, when he is hungry, or when he. is wet or cold or otherwise uncomfortable, he "reports" immediately with signs of distress. No condition outside his own body which does not impinge directly upon his sense organs is of his slightest concern. He is completely egocentric in that sense; he cannot be otherwise. When he senses discomfort he reacts immediately. He "demands" immediate relief from discomfort and immediate gratification of his needs with loud and insistent crying. He does not have to learn these patterns; they are built-in.

Babies, of course, are not all alike. They vary widely with respect to a number of temperamental tendencies. Some are by nature highly active, others are unusually quiet; some of them almost from the beginning seem to be relatively undisturbed and calm in their new conditions of existence, others appear to be displeased and uncomfortable. Babies also vary widely in their reactivity to strong stimuli: The "excitement" pattern (reactions of startle, shock and withdrawal), with its wide individual variation, is common in infants, particularly during the first month of life before the development of the inhibiting function of the brain.

Rather marked congenital differences in temperament were observed in our subjects Paul and Sally. Paul was relatively quiet, with an observant and "thoughtful" appearance, while Sally, from the beginning, was an active child, inclined toward "unevenness of disposition." She was more reactive to restrictions, with ready tears when her movements were hampered. Both babies were alert when awake and quite reactive to stimulation.

Role of Learning in Early Emotional Development

Emotional learning, however, gets underway very soon. The baby's emotional expressions begin to vary with the situation, the particular source or locus of discomfort and he begins to show differentiated expressive signs of pleasure from gratification as well as distress from pain or discomfort. As development continues with experience, he becomes responsive emotionally to a wider and wider range of stimuli. Thus, the direction of the child's emotional development and the characteristic quality of his emotional expressions are influenced by his environment and particularly by the kind of care and handling he receives. Once again, development is a product of organism-environment interaction.

The baby's responsiveness to other human beings soon becomes pronounced and, although evidence is not always clear, the consensus is that the quality of the attention and care he receives during this early period has much to do with the development of his later capacity to accept love and affection from others or to feel love for others. The evidence for this facet of emotional development comes largely from clinical observations. The development of the capacity for love is discussed in greater detail in a later section of this chapter.

Development Sequences

For an accurate assessment of developmental status in emotionality, however, just as in other functional areas, some sort of scale or sequence of stages is necessary. As is true with other generalized qualities of behaviour, no completely satisfactory series of indicators has been established, although several formulations have been proposed. Probably the best known of these is the series of stages postulated by Freud.

The Freudian Stages

The series of stages in the psychosexual development of the individual adopted by many psychologists as the basic framework of personality development are the oral, the anal, the phallic, the latency and the genital periods. Certain of these stages were briefly described in the discussion of the vital functions. We shall sketch this series of stages again specifically as an example of a sequence in the development of emotionality.

The oral period covers early infancy, when taking food through nursing at the breast and being made comfortable by this process are the totality of life. Much of the baby's experiences of release from tension and gratification come from or are associated with the oral activity of food taking.

When gratification through nursing is not forthcoming, when tension has mounted, feelings of frustration are experienced; anxiety and the sense of insecurity arise. With gratification come pleasure and a sense of security. This period, during which emotional experience—either pleasurable or otherwise centers around infant oral activity, lasts to about the end of the first year, thus, corresponding to the period of greatest dependency.

The generalized pleasure aspect of infant oral activity, quite apart from the mere relief from hunger, is emphasized by the psychoanalytically oriented students of child behaviour.

Throughout this early period the child is rapidly learning to do things for himself. He begins now to explore more widely as he masters the ability to move about on his own power. His abilities to manipulate with his hands the objects about him appear now to be replacing the pleasure he has been deriving from sucking his fingers. Furthermore, as he becomes more and more able to "get into things," restrictions are placed upon him. He begins to experience new demands and pressures. The attitudes of those around him regarding order and cleanliness begin to be felt. The job of toilet training begins in earnest. Since the eliminative functions pertain to and directly involve his own body and its sensitivities, they are "closer" to him than are any other activities. They involve tension, release from tension, frustrations and gratification. The child's preoccupation is thus, shifted from oral activity to the anal region. According to the Freudian account, the child has now reached the anal stage.

Certainly, this period in the child's life is extremely important from the standpoint of determining developmental trends and it can be a very difficult period for both the child and his parents. By the time the child has reached the age of 3 years oral pleasures, as a rule, are no longer predominant in his life. He is also well along in achieving bowel and bladder control. Concentration upon anal activities has also lessened as his manipulative and locomotor abilities and his interests in things outside himself gain precedence. His curiosity about and interest in people also are expanding. He begins to observe people, particularly their behaviour toward one another. In the typical family situation, although he takes it for granted, he is aware of and enjoys the special kind of intimacy and the love his parents show for him and for each other; in this close relationship his love for his parents deepens. He identifies with his parents. The little boy wants to be like Daddy.

According to Freudian theory, however, a change begins to take place in the child's feelings toward his mother. He becomes especially strongly attached to her and his attachment is erotic in nature. He is now entering the phallic phase of psychosexual development. Very strong "object cathexis" develops toward his mother. In his desire to have her to himself his father becomes his great obstacle. Despite his great love for his father he wishes his elimination. He is in the midst of the Oedipus complex.

This naturally becomes a very conflictful situation. He experiences "bad" wishes. He desires incest with his mother and the elimination, even death, of his father. According to theory, he may also experience deep anxiety. His father is powerful. As his rival, his father, he fears, might retaliate and rob him of his phallus. Castration anxiety may thus, develop.

In his need to escape from the stressful predicament certain "mechanisms" may come to the child's aid. Repression of his shameful, guilt-arousing wishes, along with the childish associations and memories connected with them, takes place. At the same time, he may gain some degree of satisfaction of his need for his mother by strongly identifying with his father, the one who possesses her.

Freud (1938) describes the period of infancy and early childhood as characterized by "infantile sexuality," which reaches its height during this third, or phallic, stage. He held that the apparent later loss of memory of experiences of the latency period was not "a real forgetting of infantile impressions but rather amnesia similar to that observed in neurotics of their later experiences, the nature of which consists of the memories being kept away from consciousness (repression)". He further maintained that "the influx of this [infantile] sexuality does not stop, even in this latency period, but its energy is deflected either wholly or partially from sexual utilization and conducted to other aims ... a process which merits the name sublimation."

Functional development both motor and cognitive, however, makes marked progress. A large number of locomotor and—manual skills are acquired and the child's interests and his range of information expand greatly. As a person there is steady development. During the five or six years of so-called latency a rather remarkable behavioural metamorphosis frequently takes place. The child is likely to change from a rather agreeable, compliant, talkative and generally delightful individual to a loud, intrusive and rather impudent preadolescent.

With the onset of puberal changes, the child experiences a period of physiological instability. During this fifth, or puberal period, there is much pre-occupation with the body and its changes. A step-up in interest in the opposite sex and many other changes in feelings and attitudes make this period an extremely important one developmentally. In terms of Freudian theory, pubescence sets off a reawakening, a resurgence, of sexuality that for a time has been latent.

These five periods may be taken in broad outline to represent emotional development. Theoretically, at least, they are directly related to specific stages in the maturation of the organism. The oral period is one of infant helplessness. The anal period begins as the child, through the maturation of body structures, begins to overcome his helplessness, to operate with some degree of independence and thus, to have shifted to himself some measure of responsibility for his independent acts. He must begin now to conform to the cleanliness requirements of society by gaining control of his hitherto involuntary eliminative processes. The demands of society force upon him a pre-occuption with anal functions during the period when maturation is in the process of bringing the body and those neural structures involved to a stage of development in which voluntary control is possible. The maturational basis for the early phallic period, however, is not so clear. Perhaps general maturation of the child's organ systems, particularly of the brain cortex and the neuromuscular systems, makes possible a shift from conscious efforts to control the eliminative processes and a pre-occupation with them, to an increased interest in his parents and their relationships with each other and with him.

The latency period begins with the gradual cessation of the stressful genital period. Controls have been achieved, the pressure is off and considerable independence has been gained. The child's tensions and gratifications are no longer so largely determined by the functions of the alimentary canal and his intense relations with his parents. They come now largely from a physically active life in an interesting world of things and people outside his home and family situation. The developmental changes that characterize the puberal period are quite dramatic. They furnish an obvious basis for the many changes in interests and activities, which take place during this period. Implications of puberal changes for emotional development and adjustment are profound.

A Development Sequence Based on Infant Observation

The Freudian sequence of phases in psychosexual development, as we have noted, has mainly to do with change in sources of gratification and objects of affectional attachment. For many students of personality and particularly for those engaged in clinical diagnosis and treatment, this conceptual sequence has been found useful and meaningful and has been widely accepted.

For many of the more objectively oriented students of emotional development, however, there is a need for more specific behavioural indicators of the onset and development of the various emotional behaviour patterns and affective experiences of daily life.

Perhaps the nearest approach to a useful series of developmental stages for the first two years of life comes from an investigation by Bridges (1932). In this study, based on observations of the emotional behaviour from birth of 62 children, a careful record was made in each case of a response and the environmental conditions that preceded it. One outcome of the study was a generalized description of emotional development in terms of four kinds of change, as follows:

1. Intense emotional responses become less frequent.
2. The reaction to any given stimulus becomes more specific.
3. Emotional responses are transferred from one situation to another, which did not originally evoke them.
4. Emotional responses gradually become more differentiated.

In relation to the fourth aspect of change, that is, differentiation, Bridges described, with approximate timing, the sequence that takes place from birth to 2 years of age. From the beginning certain kinds of stimulation give rise to a generalized, overall excitement, which Bridges regarded as the original emotion. Infants who are thus, excited are described in terms of suddenly tensed muscles, quickened breath, jerky, kicking leg movements and wide-open eyes gazing into the distance. This pattern was found to persist throughout the first 2 years of life.

At about 3 weeks of age the first step in differentiation takes place. Distress, or unpleasurable excitement, as distinct from generalized excitement, appears. In distress there is greater muscle tension, more interference with breathing, louder and more irregular crying at a higher pitch. Tears may flow,· the face becomes flushed, the mouth distorted and the fists clenched.

Not until the age of about 3 months does the distinctly pleasurable sort of excitement, delight, become differentiated. This pattern involves kicking, faster breathing, opening of the mouth, crooning sounds and smiling. In contrast with the distress pattern, the movements in delight are free rather than restrained, the eyes are opened rather than closed there are smiles rather than frowns and approach movements rather than withdrawal.

The next differentiations in the sequence are on the unpleasurable side. According to Bridges, at about 5 months of age the baby shows unmistakable signs of anger, as differentiated from the more generalized distress. The clearest sign of anger is a wail, as if in protest, without the closing of the eyes. Closely following anger comes disgust, characterized by coughing, sputtering and frowning while the baby is being fed. The third differentiation from distress, according to this study, is fear, observable at about 7 months of age. The presence of a stranger is the most frequent stimulus to fear. The behaviour pattern includes general inhibition of movement, tears and steady crying, with eyes tightly closed and head bent. The body remains rigid and inactive.

At about the same time (7 months) a pleasurable emotional pattern, elation, becomes distinguishable from generalized delight. The baby smiles, takes a deep breath and appears to express satisfaction in a sort of grunt.

By the eleventh month there are clear signs of affection for adults. The baby at this time will put his arms around the adult's neck and stroke or pat the face with obvious delight. Sometimes he will approximate a kiss by bringing his lips close to the adult's face. By age 15 months he makes an affectionate response to other children as well as to adults.

Very shortly after affection manifests itself, jealousy appears as a further differentiation of distress. It is characterized by tears, bent head and by standing stiff and motionless. In some instances anger and aggression are part of the picture. Finally, at about 20 to 21 months, the pattern described as joy makes its appearance.

This sequence of differentiations described by Bridges perhaps comes nearest to meeting the requirements of a series of progress indicators in emotional development. However, as we have noted, it covers only the first two years of life. The different behaviour patterns are extremely difficult to recognize and distinguish in a baby's behaviour. A given child's status in emotional development, we must therefore conclude, must be estimated mainly in terms of a general understanding of emotionality and of the changes that take place with age, rather than in terms of a definite scale of progress indicators.

Maturation and Learning

As was suggested in our discussion of the Freudian sequence, maturation as well as learning is involved in the development of emotions. Freud's psychosexual stages relate directly to bodily structures and their maturation.

Emotionality in general, however, since it is so pervasive, involving as it does the functioning, more or less, of the total organism, is not so closely bound in its changes with time to the maturation of a specific bodily structure as is the case, *for example,* with the development of mentality. The great diversity of emotional patterns and modes of emotional adjustment to be found among otherwise mature people must be accounted for very largely in terms of learning in environmental diversity. This, of course, is not to discount the fact that individual differences in constitutional nature have much to do with determining idiosyncrasies in reactions to given environmental situations.

Planned and Incidental Learning

That significant changes in the quality and patterning of emotions do take place, particularly during childhood, there is no doubt. Earlier in this chapter the two opposing categories of emotional experience were mentioned and in that connection we emphasized the importance of the prevailing quality-positive or negative—of the child's home environment in determining the general nature of his emotional and social attitudes and behaviour. More specifically, evidence suggests that the prevailing quality of the experience the child has with his mother during early childhood is of paramount importance.

I. D. Harris (1959), in an intensive study of a group of children and their mothers, concluded that a mother's loving care—her warm dependability and her understanding of her child's individual needs for expression—was "crucially" related to later desirable emotional and social adjustment and that the lack of such warmness and understanding was a factor in later maladjustment. In the words of Harris:

The mother can be a source of nutriment and pleasure for an infant as he feeds, but she can also be a source of pain, tension and frustration. If she is a source of pleasure, then the growing child will look upon others and later humans as gratifiers; if she is a source of pain and tension, the growing child will react to others and to later humans as frustrating.

During the toddler period, when the "socialization" of the child is underway in earnest, progress in learning takes place at a rapid rate. Two general types of learning influences can be identified during this period particularly. First, the child, of course, learns those patterns of behaviour—personal and social habits and ways of speaking and thinking—which the parents set out to teach him in order to make of him an acceptable member of society. But while this deliberate inculcation of the culture is taking place the child also learns things that were not deliberately planned or consciously taught by the parents, There is likely to be much incidental emotional learning taking place, *For example,* the child is likely to learn, by association, a joyful response or an indifferent one or one of dread and resentment at the appearance of his father. He may learn to react to his parents generally with positive, approaching affection, or he may learn to associate them with frustration, anger, or fear.

The sort of emotional learning that does take place, in any case, depends upon the quality of the emotional interaction that prevails between the child and his parents. And the quality of that prevailing interaction is dependent upon two sets of factors: *(1)* the particular personal make-up of each parent (predominant traits, attitudes, expectations, frustrations); and *(2)* the child's unique pattern of temperamental tendencies such as those discussed (another example of organism-environment interaction).

Imprinting versus Healthy Emotional Learning

There is another line of evidence from recent research, mainly animal research regarding the imprinting phenomenon, however, which suggests a somewhat different interpretation of the relation between learned emotional patterns and environmental stimulation. The suggestion coming out of these studies is that perhaps it is not so much the pleasant quality of the emotional interaction between parent and young child that makes for positive emotional attachment as it is that there be strong emotional content of some sort in the of some sort in the interaction during the early critical period. In Scott's words:

> It should not be surprising that many kinds of emotional reactions contribute to a social relationship. The surprising thing is that emotions which we normally consider aversive should produce the same effect as those which appear to be rewarding.... [This fact] provides an explanation for certain well-known clinical observations such as the development by neglected children of strong affection for cruel and abusive parents.

In view of apparent conflict in the evidence cited by such investigators as Scott (1962), I. D. Harris (1959) and Diamond (1957), it is clear that a distinction must be drawn between an early emotional bond due to the "imprinting process" and healthy emotional adjustment and development in children. E. H. Hess's (1960) ducklings did become the more strongly imprinted to follow the mechanical model duck the greater the effort arid the pain they encountered in following it. Harlow's (1958) infant monkeys did crawl desperately toward their rejecting, punishing mothers, who, in their own infancy, pad been reared on terrycloth dummy "mothers" instead of interactive monkey mothers. Childreri have been known to become emotionally attached to cruel, unloving parents. But Harlow's dummy-reared monkey mothers had not developed the normal monkey interactive behaviour patterns in their association with and close attachment to their dummy mothers. Furthermore, in the vast majority of instances, human individuals who have been reared by neglecting and cruel parents must overcome great handicaps in acquiring the capacity to interact warmly and affectionately with others. Scott (1962) summarized the evidence as follows:

> There are demonstrable positive mechanisms, varying from species to species, which bring young animals close to other members of their kind: The clinging 'response of young rhesus monkeys; the following response of chicks, ducklings and lambs and other herd animals; social investigation, tail wagging and playful fighting of puppies; and the visual investigation and smiling of the human infant. These are, of course accompanied by interacting responses from adults and immature members of the species: holding and clasping by primate mothers, brooding of mother hens and other birds, calling by mother sheep, investigation and play on the part of other young puppies and the various supporting and nurturing activities of human mothers.

***

5

Learning—Laws and Theory

Learning means the modification of behaviour of an individual. It is said to be equivalent to change, modification, development, improvement and adjustment. It is not confined to school learning, reading, writing, typing, cycling, cooking but it is a comprehensive term which leaves permanent effect on the individual. While approaching a burning match stick the child gets burnt and he withdraws. By this incident he learns to avoid not only the burning matchstick but also all burning things. Learning has been defined in many ways as follows:

According to Woodworth, "Any activity can be called learning so far as it develops the individual (in any respect, good or bad and makes his behaviour and experiences different from what that would otherwise have been."

According to Garden Murphy, "The term learning covers every modification in behaviour to meet environmental requirements".

According to Walker, "Learning is a relatively permanent change in performance that occurs as a result of experience."

According to Jones and Simpson, "Any change of behaviour which is result of experience and which causes people to face later situations differently may be called learning."

According to Pressey, Robinson and Horrocks, "Learning is an episode in which a motivated individual attempts to adapt his behaviour so as to succeed in a situation which he perceives as requiring action to attain a goal."

According to Henry P. Smith, "Learning is the acquisition of new behaviour or the strengthening or weakening of old behaviour as the result of experience."

According to Kingsely and Garry, "Learning is the process by which behaviour (in the broader sense) is originated or changes through practice or training."

Nature of Learning

1. Learning is the acquisition of new behaviour.
2. Learning is the total reaction of total situation.
3. Learning is the process by which an activity originates.
4. Learning is the perceptual reorganisation of the situation.
5. Learning is universal in nature. Man is rational and he learns the most.
6. Learning is goal-directed activity.
7. Learning is the change in behaviour as a result of experience.

All Animals Learn

1. Learning is motivated by adjustment.
2. Learning is a never ending growth. We always aspire to learn more and more.
3. Learning is acquisition, retention and modification of experience.
4. Learning is process and product.
5. Learning is a life long process. It starts with birth and comes to an end when the individual dies.

Main Characteristics of Learning

Yoakman and Simpson have enumerated nine general characteristics of learning which we discuss below:

1. Learning is Growth. The word growth is generally associated with the body which is growing, but through the mental growth of the learner. Although it is latent yet we can perceive its growth. Through his daily activities the child grows both mentally and physically. Therefore, we say that learning is growth through experience.

2. Learning is Organising Experience. Learning is not mere addition of knowledge. It is not mere acquisition of facts and skills through drill and repetition. It is the reorganisation of experience.

3. Learning is Adjustment. Learning helps the individual to adjust himself adequately to the new situations. Children meet with new situations which demand solution. Repeated efforts are required to react to them effectively. Life is full of experiences and each experience leaves behind some effects in the mental structure. These effects modify his behaviour.

4. Learning is Purposeful. All true learning is based on purpose. Purpose plays a big part in learning. According to **Ryburn**, "This purpose is always connected with the use of some instinctive power, with the use of the energy with which we are endowed with birth." We do not learn anything and everything that comes in our way, in a haphazard manner. All schools activities should be purposeful so that the child should feel real urge for learning.

5. Learning is Active. Learning does not take place without a purpose and self-activity. In the teaching-learning process, the activity of the learner counts more than the activity of the teacher. The principle of learning by doing is the sole principle and has been recommended by all modern educationists. It is the basis of all progressive methods of education such as the Dalton, the Project, the Montessory and the Basic.

6. Learning is Intelligent. Meaningless efforts do not produce permanent result. Any work done mechanically is without any soul. When a child learns something unintelligently, he is likely to forget it very soon. He does not assimilate but simply commits to memory. Only efforts made intelligently have lasting effects.

7. Learning is both Individual and Social. Learning is more than an individual activity. It is a social activity also. Individual mind is affected by the group mind consciously as well as unconsciously as individual is influenced by his friends, relatives, classmates, parents etc., and learns their ideas, feelings and actions. Social agencies like the family, church, film and gangs of playmates have a tremendous influence on the child and are always moulding and remoulding him.

8. Learning is a Product of the Environment. Environment plays an important part in the growth and development of the individual. Environment should be healthy and rich in educative possibilities.

9. True Learning Affects the Conduct of the Learner. There is a change in the mental structure of the learner after every experience. By Colvin learning has been described as the "modification of the reactions of an organism through experience."

Factors Affecting Learning

Learning is influenced by various conditions of life and school. Our attempts should be to create such conditions as are conductive to effective learning and which help students to make the most satisfactory adjustment to life. The following factors influence learning:

1. Hereditary Factors. Neither we can change nor increase hereditary endowment. Of course, we can use it and develop it. This differs in different individuals. Some are very rich in the matters of hereditary endowments while others are very poor. The native intelligence is different in individuals. Children vary also in particular abilities. The intelligent children can establish and see relationship very easily and more quickly. Our ability to learn and the rate of learning are conditioned by our heredity. Maximum use should be made of the hereditary endowment.

Attempts should be made to see to it that children get opportunities to use and develop their hereditary endowment.

2. Goals Set before the Pupils. A definite goal should be set before each child according to the standard expected of him. Immediate goals should be set before small children and distant goals for older ones. It must be remembered that the goals should be very clear and the children must understand these goals.

3. Physical Conditions of Children. The physical conditions of children also affect learning. Bodily weakness, chronic illness, malnutrition, fatigue and bad health are a great hindrance in learning. The home conditions—bad ventilation, unhygienic living, bad light, overcrowding etc., affect the rate of learning and the general response of the child. Sometimes children have to walk long distances to and from the school and this also influences learning.

Physical conditions of the classroom are the other determining factors. It may also be remembered that learning is affected by the physical teachers—their health, voice and speech.

4. Stimulation. Best learning takes place when the teacher is successful in arousing the interest of the students. "The guidance of the teacher is mainly a matter of giving the right kind of stimulus to help him to learn the right things in the right way", write Ryburn.

5. Association of Things. Thorndike points out that things which we want to go together should be put together as a part of one process. Then it becomes easier to make the students understand their connection.

6. Guidance and Instruction. Suitable guidance should be given to the students in selecting the best response to their environment. Demonstration is very helpful to teach various skills.

7. Result of a Total Situation. Learning is affected by the whole situation. It is not always possible to estimate the result of one individual factor. Learning is fruitful and permanent if the total situation is related to life.

8. Emotional Conditions. Children should be praised when they show good results. This gives them encouragement to show all the more better results and they develop confidence, hope, self-reliance and self-respect. Sympathetic attitude on the part of the teachers gives stimulus and a sense of security to the students. We should discard our habit of fault-finding. This develops fear and feelings of insecurity and of inferiority.

THEORIES OF LEARNING

Psychologists differ at the nature of learning process. Accordingly they have propounded different theories.

Some of the important theories of learning are as follows:

1. Learning by trial and error.
2. Learning by insight.
3. Learning by imitation and observation.
4. Learning by conditioning.

1. Learning by Trial and Error

The chief exponent of this theory was Thorndike. He make many experiments on rats and cats. He concluded that an individual makes many random efforts to reach a goal. Most of his efforts go useless in his aim. By further trial and error he goes on eliminating useless movements and synthesising useful ones. At last he reaches the aim. In further trials he need not repeat the useless movements. He will perform the activity quite easily in further attempt as he learnt to perform the real needed movements after eliminating the useful ones. His typical experience in his connection is given below:

The Cat Experiment. He shut-up hungry cat in a big cage. An attractive piece of food was placed outside the cage. The cat looked at the piece and tried to get at it. It extended its paws through the bars of the cage but the piece was far away. It tried every part of the cage but all its efforts failed. It thrust its nose through the bars but all in vain. There was a latch inside the cage. While it was making random movements its paw struck against the latch. The door of the cage opened. Next week the experiment was repeated. The cat made some random movements again but was successful in opening the door sooner than before. In the third or fourth trial there was no random movement. The cat pushed against the latch and went ouside. It learnt the method of opening the door by trial and error.

The experiment sums up the following stages in the process of learning:

(1) ***Drive.*** In the present experiment it was hunger and was intensified with the sight of the food.

(2) ***Goal.*** To get the food by getting out the box.

(3) ***Random Movements.*** **The** cat, persistently, tried to get out of the box.

(4) ***Block.*** The cat was confined in the box with a closed door.

(5) ***Selection (of proper movement).*** Gradually, the cat recognised the correct manipulation of latch. It selected the proper way of manipulating the latch out of its random movements.

(6) ***Chance Success.*** As a result of this striving and random movement the cat, by chance, succeeded in opening the door.

(7) ***Fixation.*** At last, the cat learned the proper way of opening the door by eliminating all the incorrect responses and fixing the only right responses.

Educational Utility. In solving problems of mathematics in the school and those of life in the world we do follow this method in many cases. There is a tendency in animal as well as in men to find out the solutions of difficult problems by trial and error method but the whole behaviour of man is not based on this sort of learning. Skills like walking, talking, running, swimming, cycling, reading and writing can be well explained by this theory. Pupils should remember the motto of 'practice makes a man perfect,' 'work hard' 'try, try again'.

2. Learning by Insight

Kohler and Koffka, were the chief exponents of this theory. They stressed on the totality of the process of learning. They took the process of learning as a synthetic activity which brings forth complete solutions of problems. They referred to the sudden appearance of solutions in many cases. Insight means penetration into the nature of things to find out the solution of a problem without necessary trial and error.

Experiment 1

In one of his experiments Kohler shut-up his chimpanzee, Sultan by name, in a big cage. Out placed a banana at a considerable distance of the cage. Close the cage there were put two sticks—one of them a long one and the other a bit shorter. The shorter stick could be screwed into the long one. The long stick could not reach the banana but if the other one was screwed into it, the banana could be touched. Sultan tried to get at the banana. He tried the longer stick but it did not reach the fruit. The other was still smaller. He sat down and began to play with the sticks. But he was still brooding over the matter. Suddenly an idea flashed to him. He thrust the smaller sticks into the hole of the longer one and thus, managed to get at the banana with the help of the combined sticks.

Experiment 2

In another experiment the chimpanzee was shut up in a room with unscalable walls. A banana was hanging with the ceiling.

The animal was hungry. He jumped at the fruit but it was too high. He left the efforts and sat down. There was a box lying in the corner of the room. The animal began to play with the box. He then suddenly got up and pushed the box to the centre of the room below the banana, jumped from it and got the fruit.

Kohler says that learning is a question of perception and not mere motion-activity or mere trial and error. He insists that this sort of learning is very useful in difficult and complicated institutions.

In fact, man does not always learn through trial and error. He often learns through insight. Anyhow learning by insight is much superior to learning by trial and error and we should encourage the children in this process.

Educational Implications

(a) Learning by insight leads to transfer of training.

(b) It can be used while teaching lessons because the teacher can proceed from whole to part.

(c) The teacher can arouse the motivation of pupils psychologically in the classroom.

(d) It saves time and energy because it is quick, efficient and permanent.

3. Learning by Limitation and Observation

We find that children learn much by mere imitation. They observe doing various things. They then try to intimate the same activities and learn many things in this way. Even in school the children learn reading and writing and many other processes by imitation and most of our behaviour is imitated.

Thorndike put some cats in a cage and studied their behaviour as regards the process of imitation. He came to the conclusion that cats do no imitate. Therefore, he does not give value to this theory. But the apes of Kohler imitated. In fact animals differ in this tendency. There is no doubt in it that man learns a lot by initation.

4. Learning by Conditioning

By conditioning we modify the behaviour of an individual in such a way that the response originally connected with a particular stimulus comes to be aroused by a different stimulus. The classical experiment of Pavlov will make the process of conditioning clear.

Pavlov was Russian psychologist. In one of his experiments he fitted a tube in the lower jaw of a dog. The dog was kept hungry and was offered food at a fixed time. Simultaneously with this act of offering the food to him, a bell was rung. It was natural that the dog secreted saliva when he saw the food. The saliva went into the tube and it was measured. The experiment went on for some days. After that the bell was rung one day but no food accompanied it. The dog secreted the saliva even then. It was observed that the saliva went on coming in the same quantity with the ringing of the bell for some days. The actual stimulus to bring forth the response, *i.e.,* the secretion of saliva, was the sight of the food but it was conditioned in such a way that another stimulus which ordinarily had nothing to do with secretion of saliva began to stimulate it.

Educational Implications. Most of the habits are conditioned reflexes only. Similarly, some sentiments are also conditioned reflexes. The school can exploit this mechanism to a great extent. The children learn to speak with the help of this process. Direct method of learning, the use of audio-visual aids, associating various subjects with incentives are based on this theory. It may be helpful in formulating principles regarding rewards and punishments. It is also useful in the learning of languages.

All the theories explained above help in the process of learning but man does not learn through any one of the theories exclusively. The child does learn through trial and error as the cats of Thorndike did but he does not stop there.

Often he benefits from the power of insight but he does not stop there too. He observes and imitates others also—insects, animals, men and inanimate physical forces. Many times different stimuli combine to develop conditioned reflexes in him. He synthesises all those modes of learning and goes on developing.

LEARNING

I.P. Pavlov (1849-1936), an eminent Russian physiologist and a noble prize winner for his work on physiology of digestion, while experimenting on the gastric secretion in dogs observed that any stimulus frequently associated with the presentation of food aroused salivation. This observation led to the famous conditioned response theory of learning. He noticed that an incidental stimulus was sufficient to bring salivation. Then his attention was diverted to the possibility of producing saliva to an unnatural stimulus.

Likes trial and error, conditioning deals with certain basic principles of learning process. It holds that the formation of SR connection, its strengthening or weakening depends upon association between two things.

Presentation of a stimulus arouses a response in every normal organism. When you see a friend you welcome him, when you see a snake you run away from it. Sight of food arouses salivation. Here, friends, snake and food are natural or unconditioned stimuli and responses like welcome, running away and saliva are natural or unconditioned responses.

However, when an unnatural or conditioned stimulus is frequently associated with a natural or unconditioned stimulus, the 'O' starts showing his conditioned response to the stimulus. The conditioned response here is equal to the unconditioned responses, *i.e.,* saliva. In other words, saliva is the natural response to food. If food is associated with bell for several trials, bell alone brings out saliva. In short, due the association between the unconditioned (Food) and conditioned stimulus

(Bell) the 'O' responds to the conditioned stimulus (Bells) in the same way (Saliva) as he responds to the unconditioned stimulus. This is called conditioned response learning. In case of Pavlov's dog, bell is the Conditioned Stimulus (CS) and Saliva is the Conditioned Response (CR).

Conditioning is supposed to be a process by which a response (Saliva) is attached to a stimulus (bell) different from the one which originally elicits that response. In simple words, a response is said to be conditioned when some stimulus other than the already effective one comes to arouse of modify it for the establishment of a definite stimulus response connection. A non-effective stimulus, therefore, brings an effective response due to conditioning. This pattern of behaviour is observed when learning by conditioning is complete.

The nipple in the mouth of an infant in an unconditioned stimulus and sucking is an unconditioned reflex. But when the sight of the bottle produces sucking response in the baby, it is called conditioned response. Similarly, happiness at the sight of a friend's letter, fear for the doctor, crying at the sight of water, are certain examples of conditioned responses. When an injection needle pricks the muscle of the baby he cries out of pain. Next time when only he sees the needle or the doctor, he cries with the anticipation that the needle would now prick him again. Here the pricking of the needle is the U.S. crying is the U.R., sight of the needle is the C.S. and crying is C.R. Strengthening of the SR connection therefore depends upon the association between two stimuli.

CLASSICAL CONDITIONING

The conditioning technique developed by Pavlov is famous as the classical conditioning technique. The following experiment conducted by Pavlov on a dog is known as classical conditioning experiment. Pavlov made a minor surgical operation in the cheek of the dog to measure the rate of salivation. The animal was kept in a sound-proof room. During the experiment a servant was feeding the dog. Pavlov to his great surprise noticed that one day, only the sound of the footsteps of the servant made the dog to salivate. The very incident created such an interest in Pavlov that he devoted most of his time to study the phenomenon of conditioning.

To start with the experiment, when food was placed in the mouth of the dog, it brought saliva. On other occasions, only a bell was rung, but it did not bring any salivary response from the dog.

Then Pavlov rang the bell first and after an interval of 5 seconds placed the food in the mouth of the dog. In this first trial bell alone did not bring saliva until food was not placed in its mouth. Only after the repetition of bell-food sequence for several days, when the dog could connect the bell with the food and anticipated food at the sound of bell, did it salivate to the bell only, even before seeing the food. Pavlov noticed that the dog salivated to the presentation of bell only after several trials when learning by conditioning was firmly established. From this Pavlov inferred that stimulus response connection has been established. The dog has finally learnt to respond to the C.S. This technique of establishing a response to a conditioned or artificial stimulus is called classical conditioning process. Pavlov, in course of his many experiments, demonstrated that the dog can be made to salivate to any stimulus, however, unnatural and irrelevant it may be. This discovery of Pavlov made a significant contribution to the learning theory.

Pavlov's classical conditioning can be explained by the following pattern, which Pavlov used in his experiment on dog :

1.	Natural or unconditioned stimulus (U.C.S.) food.	Natural or unconditioned response (U.C.R.) Salivation
2.	Conditioned Stimulus (C.S.)	—No salivation

3. C.S. (Bell) —Unconditioned response
   5 seconds interval U.C.S. U.C.R. (Saliva)
   (food)

Condition 3 is repeated for several trials until an association is established between bell and food (C.S. and C.R.)

4. C.S. (Bell) —C.R. (Saliva)

In any classical conditioning experiment in the C.S., U.C.S. pattern, the C.S. should always precede the U.C.S. and the time between the two must be constant and very close in all trials. If the time gap is too large! Conditioning may not be established because lack of close temporal pairing of the C.S. and U.C.S. will not be able to establish the association between C.S. and U.C.S. While establishing the salivary response to a tone in one experiement the following observations were recorded:

No. of presentations of sound and feeding	*No. of drops of saliva in 30 seconds*
1	0
9	18
15	30
31	65
41	64
51	69

[From Munn, O.N.L. (1954) Psychology page 118, Time Sequence in Conditioning.]

In this experiment after nine presentations of the C.S. and U.S. in close proximity, conditioning was established. It is also noticed from this table that the number of drops of saliva increased with number of repetitions of the conditioning process.

The strength of conditioning can be found out in terms of saliva and in terms of latency (the time elapsing between the bell and the saliva).

Time Sequence in Conditioning

Usually is conditioning technique the C.S. precedes the U.S. by a fraction of second. Sometimes it is almost simultaneous. As a matter of fact, about 3 seconds give the best result as studies show. It is also found that if we measure the effectiveness of conditioning, the strongest conditioning occurs with shorter latency period, *i.e.*, near about 3 seconds.

The most vital point in Pavlov's classical conditioning technique is that food always follows bell irrespective of the salivation of the dog. Whether the dog salivates or not, does not matter. Food is always presented after bell with the lapse of a very short time interval.

Besides Pavlov, Bechterev, Watson and many more psycho-logists have conducted experiments on conditioning. Wever made respiratory conditioning to belt. Pavlov and his associates have observed certain important phenomena in conditioning which are discussed below.

Stimulus Generalisation

It is a tendency for the CR to be aroused by a similar stimulus other than the stimulus aroused in training. Generalisation in conditioning occurs to a certain class of stimuli rather than to a specific stimulus. Pavlov noticed that when a C.S.C.R. bond has been established by conditioning, a stimulus which is similar to the C.S., can produce the same response and he called this stimulus generalisation. The more similar the new stimulus, the greater is the probability for generalisation and *vice-versa*. If the dog is conditioned to salivate to tone, it will salivate to any type of tone like electric bell, worship bell, college bell, buzzer, ding dong bell and other sounds.

Watson in an experiment on the development of fear in children has demonstrated stimulus generalisation. Albert, a small child, used to play with a rabbit without showing any fear. Watson produce a loud sound whenever the child started playing with the rabbit. After a number of trials, the child started showing fear towards the rabbit, associating rabbit with the occurrence of loud sound. Then the child showed fear in gerenal towards while furcoat, bunch of cotton and to all other stimuli having similarity with white rabbit. Stimulus generalization in conditioning happens usually more in childhood particularly when the child has not developed the capacity of differentiate between two stimuli. This is why, during infancy the baby considers every women to be his mother. This tendency of generalization Pavlov attributed to a spread of effects from the region stimulated to other parts of the organism.

Stimulus Differentiation

Differentiation develops out of generalisation. Though in the beginning mostly generalisation in conditioning is found, the organism can be conditioned to make a response only to a specific stimulus by appropriated conditioning procedures.

Suppose you want to positively condition a dog to a tone of 400 cycles and negatively condition to 800 or 1200 cycles. You present the reinforcement only after 400 cycles tone and do not present reinforcement when the other tones are presented. If this is done for several trials the dog will learn to discriminate between these tones eventually. In this case the animal is positively conditioned to 400 cycles and negatively conditioned to 800 and 1200 cycles. Since it is not given any reinforcement at 800 or 1200 tones, these are not reinforced.

Children gradually learn to differentiate one stimulus for another in their own environment. In the early period they consider any woman as their mother. But gradually, as they grow up, they differentiate their mother from other woman. The woman who fondles, nurses, feeds and takes every care of the child and sleeps with him is considered as his mother.

Discrimination in conditioning has certain utilities in learning. Though usually stimulus generalization facilitates learning making, it is easy and quick, sometimes it also inhibits learning, creates confusions misunderstanding and many emotional problems. Like Albert, if one is afraid of any stimulus which has some similarity with white rabbit, he will have many adjustment problems. Differentiation at this point can prove useful. It can overcome the interfering and inhibitory effect of stimulus generalization. At the same time, if the differentiation is pushed too far, it leads to experimental neuroses. Munn, therefore, remarks "Sometimes as the difference is reinforced and unreinforced, stimulation becomes too small for the animal to discriminate, a nervous breakdown occurs."

Experimental Extinction

Once a C.R. is set-up, if one goes on repeating it without giving reinforcement, it tends to be extinguished. Sometimes people ask, how can one eliminate a conditioned response once it is established? The answer to this is experimental extinction. The process by which a C.R. can be weakened and finally eliminated is called the process of experimental extinction. It is a process by which a well established C.R. is eradicated by repeating the C.S. without the application of reinforcement. Take the case of Pavlov's dog. After the dog has learnt to salivate to the sound of the bell, it will continue to salivate to the bell for a certain period, although food is not given after bell. However, after some such trials when the dog will see that no food is coming after bell, the connection between bell and food will grow weaker, the flow of saliva will decrease. At last, the dog will not salivate to the bell at all.

Similarly, in a study of "The Eradication of a tactile conditioned reflex of man the shoulder region of the subject was conditioned so that is elicited a galvanic skin reflex. Vibration of this region was then repeated without presentation of shock. There was a gradual weakening of response and finally complete elimination".

Internal Inhibition

After the establishment of a C.R. if the reinforcement is not provided check to salivate becomes stronger and stronger. So much so that it becomes inter-natural in nature.

Internal inhibition is a kind of negative learning not to salivate and can be applied to Ebbinghaus's curve of forgetting. But when the inhibition becomes weaker it leads to spontaneous recovery like reminiscence in forgetting. The idea of Ebbinghaus's curve of forgetting was applied to the concept of internal inhibition. It was said that this check or block was something which is acquired, *i.e.*, learning not to salivate. The negative learning not to forget is also forgotton to the same extent. However, this extinction is only a temporary suspension of response because of an internal set.

Spontaneous Recovery

Pavlov was of opinion that complete or permanent extinction of a C.R. is not possible. When the experimentally extinguished response reappears again after a period, it is called spontaneous recovery. This concept can be compared with reminiscence in forgetting. In this process even without the presentation of food, bell produces a few drops of saliva. Pavlov noticed that once the flow of saliva was completely extinguished, after about 30 minutes of rest interval the dog again salivated to the C.S. without U.S. But if again no reinforcement is provided after the C.S., spontaneous recovery will grow weaker and weaker and would at last fail to reappear.

Higher Order Conditioning

In course of his experiments on dog, Pavlov found that once conditioning was firmly established, for further conditioning, he could use the conditioned stimulus as unconditioned stimulus. When the dog is conditioned to salivate to Bell, Pavlov used this Bell (instead of food) as a U.S. for establishing further conditioning. Pavlov was of opinion that a conditioned stimulus can be used in much the same way as an unconditioned stimulus. He stated that more elaborate forms of behaviour can be developed by the process of chain of conditioned response.

First order or original conditioning

Food	—	Salivation
Bell	—	?
Bell	—	Salivation

Second order conditioning

Bell	—	Salivation
Light	—	?
Light	—	Salivation

Third order conditioning

Light	—	Salivation
Colour	—	?
Colour	—	Salivation

The arousal of the same response, *i.e.*, saliva to light by frequently associating it with bell but without presenting the food is called second order conditioning, as it could conditions a second neutral stimulus.

Though Pavlov claimed that a third order conditioning can be possible in the same way he could not proceed after the second order in case animals. However, with human beings, all sorts of complex learning can be possible by the process of higher order conditioning, he asserted.

Negative Conditioning/Unconditioning

Otherwise known as backward conditioning or negative adaptation it is learning not to make a response. The 'O' by the technique of backward conditioning can be made not to show a response to a neutral stimulus. Thus, instead of U.S. (Food) following the C.S. (Bell), C.S. will follow the U.S., *i.e.*, reverse to the pattern of classical conditioning technique, food (U.S.) will be given first and Bell (C.S.) will be presented next with an interval of 5 to 6 seconds between each presentation. As the 'O' will get the reward first it will not have any motivation or need to make a response (to salivate). So the C.R. would not be strengthened. If at all there will be any salivation, it will be marginal.

Many of our bad habits like nail biting, bed wetting, moving the leg all the while, various ticks and mannerisms, thumb sucking, smoking, alcoholism, breast feeding in case of older children, irrelevant fears can be removed, withdrawn by negative conditioning. Various habit patterns can be broken by this procedure. Thus, negative conditioning has valuable implications in practical life.

Whenever the rat proceeds for food, if it get electric shock, it will avoid food. A child who is punished while entering into a river will avoid the river in future. Mann reports that is sexually mature male rat is given electric shock everytime is approaches a female rat, it soon avoids female rats. When cockroaches are given electric shock for running into the dark, they learn to go to the lighted place instead of approaching for darkness. Tigers of Sunderban (West Bengal) over the last decade have killed at least 650 persons officially and the unofficial record is about 1000. The forest department worried over the increase in human deaths introduced something which was new in India, the use of electrified dummies. These dummies which look extremely life-like have been placed at a few corners of the sanctuary. Officials explain that when the tiger attacks the dummies, it gets electric shock and this persuades it from further attacks on humans. The projects officials claim atleast an 80% reduction in the death rate due to this method of withdrawal conditioning.

When alcoholics are given to take alcohol containing a drug which produces violent vomiting, they develop aversion for alcohol. Similarly, for babies who do not give up breast feeding, quinine is smeared on the nipples and then only they develop aversion for breast feeding. Negative conditioning is, therefore, known as aversive conditioning.

Withdrawal Conditioning

Becheterev, another Russian physiologist and a follower of Pavlov, taking animal and human subjects, found that when the 'O' gets a painful stimulation in a part of his body, it withdraws from it. Withdrawal may take place from electric shock, beating or any painful stimulation. Bell or light is presented before shock. After some trials by seeing the light or hearing the sound of the bell the 'O' withdraws that specific part of the body where shock is anticipated.

Experimental Neuroses

In recent times, the phenomena of neuroses has been studied by the process of generalisation and differentiation. If differentiation is pushed too far the 'O' fails to discriminate between two stimuli, becomes completely non-adoptive and is unable to salivate to any stimulus due to conflict. Consequently, nervous break down occurs. The animal suddenly becomes wild. It resents the situation.

An experiment on circle and ellipse can be referred to demonstrate experimental neuroses. When the circle and ellipse became very much similar, the animal unable to discriminate became wild. Physiological changes in respiration, heart beat, blood circulation etc., took place. It showed abnormal symptoms and behaviour disorders.

In experimental neuroses there is a clash between positive and negative forces to or not to salivate. Anxiety neuroses is produced experimentally by this method.

Implications of Conditioning in Practical Life

The contribution of Pavlov to the psychology of learning can be said to be equally important like Freud's as Pavlov as advanced certain concepts like generalisation, differentiation, withdrawal conditionings, unconditioning and experimental neuroses which have tremendous significance in life. In fact, Pavlov's discovery was so powerful that it became completely Pavlovian.

Conditioning works in every sphere of life. Bulk of learning in animals and children develops due to conditioning, language development and different habits, take place because of conditioning. Therefore, it became a key technique not only for physiology, but also for psychology, mental hygiene and therapy. Watson, therefore, remarked that the whole system oif behaviourism he has worked out in terms of habit.

When Watson found that habit is not as simple as he thought, he realized the role of conditioning in habit development. Thus, he remarked, "I had worked the thing out in terms of habit formation. It was the enormous contribution Pavlov has made and how easily the conditioned response could be looked upon as the unit of what we had been calling as habit. I certainly form that point on gave the master his due credit."

Observation of animal and human behaviour led Pavlov and Watson to believe that all our habit patterns, behaviours, attitudes, likes and dislikes, values, etc., can be attributed to conditioning. Conditioning also works as a technique for behaviour modification, elimination of behaviour problems. It is also very significant in emotional learning. Watson and Rayner (1920) believed that it is the only way by which emotions can be acquired. "Since fear or anxiety is a common symptom of mental disturbance we are led correctly to anticipate that classical conditioning figures in the development of such disorders".

Thus, the sort of dream which Pavlov had, came true to a large extent. Due to the practical implications of conditioning in life it has occupied a very prominent position in the learning psychology even today occupied a very prominent position in the learning psychology even today.

LAWS OF LEARNING

Thorndike formulated laws of learning. Laws of learning are divided under two parts:

(1) Primary law, (2) Secondary law.

Primary law of learning

Law of Readiness   Law of Exercise   Law of Effect

1. Law of Readiness. Readiness means a preparation for action. It is a kind of 'action tendency' aroused through preparatory adjustments, sets and attitudes. The fulfilment of this tendency is satisfying and its non-fulfilment is annoying. Thorndike employed physiological terms in describing this law.

Learning is a matter of interest. The learner must be made ready to learn. When a person feels ready to learn, he learns more effectively and with greater satisfaction than when not ready. There are some motives, desire or purpose without which no learning is possible. So, for learning the learner must be mentally ready.

Educational Implications. This law can be effective in the following ways:

(a) To arouse the interest and attention of the students by asking suitable questions.

(b) This law is useful in arousing curiosity of the students. Curiosity is indispensable for learning.

Motivation is an important factor for learning. It is also based on the principle of readiness. If the learner is properly motivated he will be ready for the work and apply his energies vigorously.

2. Law of Exercise. This law is known as law of practice or law of use and disuse. If any activity is exercised we come to have efficiency in doing it. If it is not repeated we soon forget it. It means repetition of an activity strengthens the nerve connections and on the other hand disuse weakens the bond. In short it can be said that law of exercise as a whole emphasises the need of repetition practice and drill work in the process of learning. Children learn spellings and pronounciation according to this law.

Educational Implications. In educational field this law can be effective in the following ways:

It Leads to Formation of Habits. If good acts are repeated again and again then they become habit. So it is the duty of the teacher to see that children are given a great deal of practice in good activities.

Correct Pronunciation. It has to be learnt by practice and incorrect pronunciation can be unlearnt only by avoiding its repetition.

By Practice Handwriting can be improved. Bad handwriting can be improved by practice. For this the teacher can ask the students to do regular practice.

Bad Habits can be Eradicated. According to Dunlop Bata theory of learning, if the children are made to practice consciously, then they automatically tend to leave their bad habits. *For example,* if a child is in the habit of thumb-sucking or nil-biting (chewing) then he should be asked by the teacher not to repeat this habit consciously. The underlying idea is that he will become conscious of this mistake and avoid the same in future.

The element of forgetfulness can be delayed or avoided because by practice the activities become deep rooted. So forgetting can be avoided to some extent.

Skills like music, painting, typing shorthand, athletics etc., can be developed to the maximum with the help of this law.

Mathematics is dependent wholly on practice. Arithmatical tables, formulas and definitions can be learnt only with the help of practice.

3. Law of Effect. Thorndike has defined this law as—"When a modifiable connection between a situation and response is made and is accompanied by a satisfying state of affairs, that connection's strength is increased, but when made and accompanied by an annoying state of affairs its strength is decreased."

It is also called as the Law of Pleasure and Pain or Law of Satisfaction and Annoyance. The word effect means the effective result of the activity. The child has to perform a number of activities. Every activity has its effect. Some activities end in pleasure and others end in pain. The activities which are associated with pleasure or satisfaction tend to be repeated again and again and the learning becomes more effective. Activities which are assoicated with pain, punishment or annoyance are not repeated and their strength is decreased. The entire system of rewards and punishments is based on this law.

Educational Implications. The educational utility of this law is as follows:

Problem Behaviour and Delinquent Behaviour can be improved by associating it with annoying state of affairs.

Theory of reward and punishment is based on this law. Reward should be given for desirable behaviour and punishment should be associated with undesirable behaviour. This law is helpful in forming and developing the desirable sentiments in the pupils. Positive sentiments should be associated with satisfying state of affairs.

Formation of Good Habits. Good habits can be formed in children by associating them with reward, satisfaction or praise. Undesirable habits can be eradicated by associating them with punishment, dissatisfaction and blame.

Memory is also connected with the law of effect. Pleasant things are generally remembered better than unpleasant things.

Interest is directly related to this law. Interest causes satisfaction and satisfaction promotes learning and better learning gives higher satisfaction. In the selection of subjects, books, hobbies games and curricular activities pupils prefer which is interesting to them.

Control of Emotions. With its help, the teacher can control the emotions of anger and jealousy of the students.

Some times it appears that a punishment is more effective than reward. Strong punishment brings about a desired attitude towards learning.

It is also found that it is human tendency to remember certain highly unpleasant incidents better than pleasant incidents.

Secondary Laws of Learning

Secondary laws of learning are as follows:

Law of Multiple Response. This law implies that when an individual is confronted with a new situation he responds in a variety of ways before arriving at correct response.

Law of Analogy. An individual responds to a new situation on the basis of the responses made by him in similar situations in the past. He makes responses by comparison or analogy.

Law of Associative Shifting. This law is also known as the conditioned response. A response may be shifted from one situation to another which is presented at the same time. In other words any response which is possible can be linked with any stimulus.

Law of Partial Activity. According to this law the learner has the capacity to select the important from the irrelevant element in order to determine appropriate responses.

Law of Recency. It means that whatever is recently learnt by the child is retained by him. The law may be defined as 'recent acts are lasting'. Hence, students should revise their courses just before the examination.

Law of Sets and Attitudes. Learning is guided by a total attitude or 'set' of the organism. The learner performs the task properly if he has developed a healthy attitude towards the task.

Law of Intensity of Stimulus. To this law, if a stimulus is strong the response will be strong and *vice a versa*. Examinations present an intense stimulus to study and hence, bear a positive effect on learning.

Law of Belongingness. This law shows that if a response belongs to a situation, the learning will be more effective. Connection between stimulus and response is natural. The teacher should create natural atmosphere in the classroom and appeal to the natural tendencies of the child.

Law of Primacy. Learning that takes place in the beginning is the best and lasting. It is a normal saying that 'First impression is the Last impression'. It implies that pupils should make the right start and should be serious from the first day. Teacher should impress his pupils from the first day of teaching. It seems the experiences that are acquired by the child at the primary stage have lasting effect.

Factors or Variables of Learning

Maturation. Learning depends upon maturation. If the individual is matured to learn a particular activity, he will learn effectively. If the individual is no mature, learning will not be effective.

Readiness. If the learner is ready to learn a particular activity he will learn better and quickly.

Capacity. The greater the capacity for learning a person has, the better will be the learning.

Attention and Interest. If the learner is motivated to learn a particular task he will take more interest in the task with full attention and hence, will learn that task better.

Memory. A learner who has good momory will learn quickly and effectively.

Motivation. Motives like reward, completion, level of aspiration, success and punishment etc., are powerful incentives to the learner for better learning.

Mental Health. A child who is mentally healthy *i.e.,* free from frustrations, conflicts, anxieties and worries will learn better.

Perception and Sensation. Perception and sensation are the basis of all cognitive learning. The stronger the power of perception, the greater the amount of learning.

Food and Drugs. Food also affect learning. Poor diet, alchoholic drugs, tobacco and addictive items adversely affect learning.

Fatigue. Fatigue causes boredom and indolence and hence, has negative effect on learning.

Will. Will to learn is an important factor of learning and it arouses, sustains and regulates the pattern of activity and leads to better attention, better motivation and hence, better learning.

Practice. 'Practice makes a man perfect' is a well known proverb. Learning is more efficient when practice is distributed at intervals over a period of time when it is considered in one period as in cramming.

Time of Learning. Experiments have shown that there are significant variations in learning efficiency during different hours of the day. It has been established that morning and evening hours are the best hours for study. During the day there is decline in the mental capacity due to noise.

Timely Testing. Occasional and periodical testing motivates the pupils to be regular in their studies. If the pupils are regular in studies they can learn better. Moreover through tests the learner knows his exact achievement, his accomplishment and there is little scope for under estimation.

Guidance. Guided learning is always better than unguided learning, Guidance saves times and energy of the learner, eliminates wastage and stagnation. But too much guidance should be avoided so that student may not learn the habit of remaining dependant upon teachers.

Knowledge of Result. Knowledge of result particularly the specific knowlege of progress of the learner improves his performance, motivates him for greater learning and serves as an incentive towards increased effort.

Vigorous Application. Slow learning is seldom efficient. Learning vigorously is a great asset and it pays much. The students should be taught to work whole heartedly. They must put their heart and soul while learning something.

Pleasantness of Task. Generally pleasant tasks are learnt quickly than the unpleasant tasks.

KOHLER'S INSIGHT LEARNING THEORY

According to German Getalt psychology Kohlers learning is insight. Insight is a important constituent in the solution of problems and is found in the higher class of animals and human beings. It is the best method among the methods of learning.

According to Kohler a person can deduce the solution by insight if he perceives the situation as a whole. Kohler, prepared some simple problems for experiments with dogs, hens, monkeys and chimpanzees. In an experiment a hungry animal was released from the house while some food was placed behind the fence adjoining the wall. Both the dog and the hen trotted around in the vicinity of the wall but as soon as they found the way out they made their exist and reached the food. This perception of the change in the meaning of the wall in insight. Bofore the insight the wall was an insuperable obstacle, but after insight it was not longer an obstacle but an object necessitating of new relations, the discovery of new patterns and the formation of new organisations.

The most famous experiments conducted by Kohler in relation to insight were those that were carried out on chimpanzees. Some bananas were placed outside the cage of a chimp called Sultan, who was then given two sticks so cinstucted that they could be fitted together. Sultan tried to pull the bananas with the sticks, an effort which he kept up for an hour, but he got tired of the attempt and gave it up for playing. While playing he brought the two ends together and suddenly he had an idea which resulted in his fitting the smaller stick in the hole of the bigger one. He then used the two together to draw the banana inward. The two sticks disintegrated but he fitted them together again. He then pulled in everything within reach as if he were trying out his discovery. The next day he took far less time to fit the two together. It is a peculiarity of insight that once the solution is learnt, it is not forgotten though its memory may become hazy with the passage of time.

The above experiments make it quite obvious that learning by insight has certain characteristics of its own. They are briefly as follows:

1. Insight is sudden.
2. Insight alters perception.
3. Insight is relative to the intellectual level. The higher species of animals including human beings have more insight than the members of lower species.
4. Old objects appears in new patterns and organisation by virtue of insight.
5. In insight understanding is more useful than dexterity of hands.
6. Some psychologists ray that learning by insight is associative learning. Insight appears suddenly after the manipulation of thought or objects for a small thought significant length of time.
7. Previous experience is of assistance in sight, though its excess does not necessarily increase insight because organised perception too is an essential factor in learning.

8. If the pieces essential for the solution of the puzzle and present together when perceived, insight comes about earlier.
9. Maturity also affects insight as evidenced by the smoother working of insight in older age than in adolescence.
10. In Woodworth's opinion, insight is sometimes hindsight and at others it is foresight. The quote him, 'Foresight is seeing the way to the goal before taking it our perceiving the uselessness of a certain lead without trying it and hind-sight is observing that a lead is good or bad after trying it. When the whole situation is clear and above board, there is a good chance for foresight, but when important characteristics of the situation have to be discovered by exploration and manipulation, hindsight is the best we can expect.

The above-mentioned characteristics of learning through insight apply both the human and animal learning. The solitary difference is the fact that a human being by virtue of his superior capacities of experiencing and understanding; observes relation between various objects quickly and sees the patterns concealed in them with alacrity. He is assisted in this by language. As a matter of fact insight occurs only when the learner perceives the related link hidden in the activity. The learning curve is altered suddenly due to insight. Insight is impossible in an extremely unfamiliar problem, because familiarity is very necessary for insight. A student of literature will have no insight in any scientific problem because he does not know even the a b c of science. When the subject of study becomes a part of life, insight becomes easier. History can be made alive by exhibitions, Geography by travelling and Arithmetic by shop accounts. Fimiliarity is also important besides the theoretical study.

One example of human learning by insight will suffice. A student of Arithmetic tries hard to solve a difficult problem of Arithmetic. He tries many methods but fails to secure a solution. He gives up and lays it aside or goes to sleep. After some time suddenly or when he is awakened the solution suggests itself and he notes it down in his notebook. When he is confronted with similar problems in the future, he will have no difficulty in solving them. The insight method is superior to imitation or conditioned response because both the later methods take more time. It is superior to trial and error in the following respects:

1. Trial and error emphasises the acquisition of motor skill but in insight mental effort is stressed upon.
2. While insight depends upon perception, trial and error depends upon sensory motor coordination.
3. Every new problem has to be tackled from the very beginning if the trial and error method is used. On the other hand, both generalisation and differentiation are present in insight, which adds to the possibilities of transfer of learning and the use of old insight in the solution of novel problems.
4. The person keeps an eye on the goal in trial and error and all activity is goal directed, but in insight it is the unconscious mind which exerts the most while the conscious activities are either very few or aimless.

Students will be permanently helped by the use of the insight method of learning in place of the trial and error method in their study and other daily activities. They will also be above to conserve their energy. Suppose that a person has to arrange furniture in a room. An average person will determine the position of the table by using the trial and error method and shifting the table from place-to-place. An intelligent person will take up a position from where the whole room in visible and imagine all the possible positions of the pieces of furniture in an imagined framework of the room and then place the respective pieces accordingly. But this does not mean that trial and error method is not meant for the intelligent. But no means is this so. In many noval situations progress is not possible without it.

THORNDIKE'S TRIAL AND ERROR THEORY OF LEARNING

Trial and Error Theory of Learning. For nearly half a century, one learning theory dominated all others in the U.S.A. despite many attacks on it and the rise of its many rivals. It is the theory of E.L. Thorndike (1874-1949). This theory was given by him in 1898, in his book the *'Animal Intelligence'*.

The basis of learning accepted by Thorndike in his earliest writings was association between sense impressions and impulses to action. Such an association comes to be known as a 'bond' or a 'connection'. Because it is these bonds or connections which become strengthened or weakened in making and breaking of habits. Thorndike's system sometimes has been called a 'bond' psychology or simply 'connectionism'. As such it is the original S—R Psychology of learning.

To quote K.P. Pandey—Thorndike's theory of learning is based on the following three elements:

(a) The Stimulus element which involves an environmental event.

(b) The formation of a connection or bond which implies that every stimulus is linked with some kind of response and thus, makes for the building block or structure of behaviour. Pandey in his book "*Advanced Educational Psychology for Teachers*" further states that the *formation of a bond* (S-R bond) is the key expression in describing the learning theory of Thorndike. When an oganism is said to have learned something, it is essentially the formation of a connection or bond between a stimulus and a response.

(*c*) The response element which indicates a behavioural act.

Connectionism has two phases:

First phase was before 1930.

Second phase was after 1930 because Thorndike made important changes later on the basis of experiments. We shall examine this theory as follows:

There are four characteristic features of Trial and Error Learning :

1. There is some sort of a motive that arouses or sustains the activity. This motive may appear in the form of need, a problem or a goal. Essentially the equilibrium of the organism is disturbed and this impels the organism to react to the situation in an effort that comes to terms with it.
2. Several different kinds of responses are made.
3. There is progressive elimination of superfluous unsuccessful or wrong form of activity.
4. Finally there is a progressive integration and establishment of the reaction by which the goal is achieved.

The essential point in trial and error is the second point—the process by which the correct response is identified through trying various responses until a solution which permits the attainment of goal is achieved.

The Puzzle Box Experiment. Thorndike studies the character of Trial and Error learning in a number of experiments on cats. A favourite device was his puzzle box used years ago by him for his studies of animal learning.

In one of the experiments a hungry cat was placed in the box and the door was closed. A fish was placed outside the box. The cat was given 100 trials, ten each morning and each afternoon for five days. The cat was fed at the end of each experimental period and then was given nothing more to eat until after the next session. In after opening the door in any trial by jumping or by chance, he went in immediately to the food, he was allowed a small bite. A complete record was made of the cat's behaviour during each trial.

It may be mentioned that besides these cat's experiments which became so common and cheap, it has trained some classical experiments performed by Ebbinghaus on memory and by Brain and Harter on Telegraphy. But these experiments of Thorndike are different from Ebbinghaus experiments w.r.t. two points and those points are that of : *(i)* Motivation, *(ii)* of Reward and punishment. He aid that the world cannot ignore reward and punishment and motivational factors.

As early as 1913, Thorndike wrote, "Both theory and practice need emphatic and frequent reminders that man's learning is fundamentally the action of laws of readiness, exercise and effect. He is first of all an associative mechanism working to avoid what disturbs the life processes of the neurons. If we begin by fabricating imaginary powers and the faculties or if we avoid thought by loose and empty terms or if we stay lost in wonder at extraordinary versatility and inventiveness of the higher forms of learning, we shall never understand man's progress or control of his education."

A serious limitation of this theory is its lack of understanding. Thorndike did not consider understanding important. He merely assumed that it would follow as a natural result of well organised learning. He believed that insightful learning, though it did occur, was infrequent.

Educational Implications of Trial and Error Theory

1. Effect of Motivation. It is the most important factor in all learning. Motivate the child.

2. Exercise Patience. Learning by Trial and Error theory is always gradual. The teacher must exercise patience with the child if the child does not show any progress immediately in any learning situation.

3. Place of Exercise. Exercise has its own place in education and learning. Repeat, give a sufficient drill in some tests.

4. Make Learning Meaningful. Effect of belonging to a thing should be exploited.

5. This theory has drawn our attention to the principle of learning by doing. You will learn only when you do a thing.

6. We must learn basis things needed for it in any learning.

S-R BOND THEORY OF THORNDIKE

One of the most accredited among all the current theories on the law of learning is the one presented by Thorndike, in America. Talking of its importance Tolman says, "The psychology of animal learning, not to mention that of child learning has been and still is primarily a matter of agreeing or disagreeing with Thorndike or trying in minor ways to improve up him. Gestalt psychologists, conditioned reflex psychologists, sign gestalt psychologists, all of us here in America seem to have taken Thorndike, overtly or covertly, as our starting point."

Thorndike first presented his theory in his book '*Animal Learning*' published in 1968. The clearest form of his theory is seen in the three volumed treatise '*Educational Psychology*'. His theory may be understood in a systematic way with the help of three major laws which are as follows:

1. Law of Readiness.
2. Law of Effect.
3. Law of Exercise.

Writing on the subject of the importance of his laws in the action of learning Thorndike says, "Both theory and practice need emphatic and frequent reminders that man's learning is frequently the action of the laws of readiness, exercise and effect." Accordingly, in Thorndike's opinion, man's learning takes place according of these laws.

He drew a learning curve for men and animals and concluded from it that usually the methods employed are similar in both classes of creatures. Until 1930, Thorndike laid particular emphasis on these three laws.

1. Law of Readiness

The law of readiness describes those situations in which the person who learns either invites the object of his learning or rejects it. Readiness include all those preparatory adjustments which immediately precede the action. Reminding the learning of his past experience mental preparation for the understanding of new things, diverting his attention towards the subject to be learned, the altering of the environment to suit the learning, are all included in learning. This readiness creates a desire for learning and turns the learners's mental attitudes towards the suject to be learned.

In Thorndike's the view law of readiness is active in three following conditions:

1. When a conducting unit is prepared to go into action, its work is quite satisfactory because nothing is done to alter its working.
2. The inactivity of a conduction unit which is ready to behave, may be unsatisfactory and any reaction may arise is connection with that deficiency.
3. When a conduction unit is forced to act while it is not prepared to do so its behaviour is of a nature calculated to excite anger.

2. Law of Effect

The meaning of the law of effect is the effect of learning in answer to question as the success or failure of learning has very significant bearing on the law now under study. Success brings with it satisfaction and along with it a strengthening of the relation of the facts. Failure increases dissatisfaction and the absence of the relation amount the facts weakness them. Thus, we may compare success to a reward or failure to a punishment and desire to repeat success or avoid failure or the inevitable antecedants.

In 1932, Thorndike propounded a new theory on the basis of his experiments about the law of effect and this he entitled "Spread of effect." This novel theory postulates that by the activity of learning adjustments are not left unaffected. *For example,* the son of a businessman is ahead of his class in arithmetic because he has already learnt some accountancy in connection with sales in the shop.

In 1935, Thorndike made some modification in the law of effect, based on experimental findings. He now said the influence of reward or punishment is not equivalent and neither it is paradoxical. The effect of reward is for more influential than that of punishment, the strengthening of associations due to reward is not paralleled by the demoralisation caused by dissatisfaction as possible in order that he may learn with success . Encourage the child when he does some thing and congratulate him besides rewarding him with object of desires in order to bring desirable spirit in his learning.

3. Law of Exercise

The second law, of any consequence in human and animal learning is the law of exercise, which is based on the laws of use and disuse. The repeated application of an activity fixes it firmly in the mind while on the other hand no psychological reference is intended. The relation is weekened through continuous disuse. Whenever there is an appropriate situation, the activity which is firmly entrenched might take place. This law, taking a cue from this universal experience, emphasises repetition in the activity of learning. Repetitive application of the activity results in the formulation of a habit in the muscles and the nerve fibres of the brain so that there is a facility in its execution in time of need. The element of understanding in repetition was not adequately stressed in Thorndike's law of readiness. *For example,* it is a rule that if repetition in the memorising of a poem is complemented by an attempt to understand and mutual relate the different stanzas, the learning may be affected with ease. Not so with Thorndike, who considered the mechanical repetition of learning individually sufficient. This type of memorising becomes rote learning highly undesirable thing and the forgetting of one line resuls in the forgetting of the rest while the recollection of the important makes a clear understanding of the subject-matter and constant renewal quite as important an constant repetition. Thus, Thorndike's mechanical repetition came in for a lot of criticism and he reached by effecting modifications.

Importance of the Major Law

These major laws enunciated by Thorndike are of great practical importance. Very few of us realise the effect which reward, encouragement, praise etc., have on learning. These theories are of importance not only in the teaching of others but also in personal activity to be learned should be created in the learner, be it oneself or some other, because the learning will be greatly expedited. And the meaningful practice of the learnt activity will help its retention in no small measure.

Subordinate Laws

Other five subordinate laws were also mentioned by Thorndike in relation to the three major ones studied above. Though a methodical explaination of true laws as given by Thorndike his later expositions sometimes contain their description while at other occasions they are lacking in it. He does not consider these subordinate laws to be of equal importance with the three major ones mentioned above. These five subordinate laws are:

1. Multiple Response. The first subordinate law is called varied reaction. A person seeks to solve a novel problem placed in front of his in accordance with his instincts and learned behaviour, because his search in conditioned by these two. There are many multiform and diversified responses in behaviour which guide a person in attaining success and makes learning possible. The learner reaches the solution by the variation in the response.

If the activities are not multiform the person would never get the correct solution. The theory of learning by trial and error if founded on multipurpose response and diversified reactions. If a person adopts a particular method of approach but does not change it in case of failure and gives up his attempts or if nothing new suggests, that person will never learn anything.

2. Prepotency of Elements. The fourth subordinate law of learning is the prepotency of elements. Every problem has some fundamental proponent element. The learner gets greater success if he bases his response on those elements, though this depends on the insight and capability of the learner. This kind of capacity is of great help in though learning. The insight, into a mathematical problems, of a student of this particular subject cannot be equalled by that of a student of languages into the same problem because the student of mathematics will grasp its fundmentals with consummate ease.

3. Set or Attitude. This is the second law of learning. Learning is conditioned by the attitude and set of the creature. The response of a person in a specific situation is dependent upon his permanent adjustment in a particular culture. On seeing a figure or idol a Hindu child will bow his head but a Muslim will not pay any respect. Both these responses are affected simultaneously by current tendencies. In an extraordinary situation, the worshipper may even throw the status. The set or attitude decides not only the response of the person but also the object from which he will derive satisfaction or classification. A first class student cries on scoring 50 percent marks, while a third class student feels delighted with the same score. The reaction of a student on scoring certain number of marks depends upon the set, attitude and the tendency present in him at the time.

4. Response by Analogy. This is the third law of learning and is called the law of assimilation. A person will respond in a manner similar to the one in which he has behaved previously if he finds that similarities between the previous situation and the new situations are none too numerous, the response will be similar of that in the first.

5. Associative Shifting. This is the first and last law of learning. The title quite obviously indicates that the position of the response of the learner shifts. This shifting is done in respect of the basic stimulus or some related associative stimulus. To take an example, a dog starts salivating when he sees food and approaches the bearer with his tail wagging. The result of constant repetition of this phenomenon is that dog wags his tail on the mere perception of the man and the shape of dish. Thus, the response originally directed to the food is now aimed at the man and the dish because both these objects are associates of the main stimulus, the food. Thus, an associative shifting in the response is observable. All the law subsidiary and main, presented by Thorndike are important in learning, both for the teacher and the student. They are applied comprehensively in education though all the psychologists are not agreed upon the truth and usefulness usually attributed to them.

The above described laws of learning are not intended to convey the impression that Thorndike's attention was confined only to the laws. Actually, Thorndike paid even greater attention to the practical aspect education and made attempt to absolve the difficulties to the teachers and taught alike. Thus, Thorndike attended to the specific aspects of learning and did not rest content with a general description of the activity. He has mentioned many factors which may assist in learning but which are not implied in and cannot be derived from his laws of learning. *For example*, he has stressed the following five things which assist in the progress of learning:

(*a*) Interest in the work.

(*b*) Interest in improvement.

(*c*) Problem Attitude.

(*d*) Significance.

(*d*) Attentiveness.

To these five. Thorndike added two which in his opinion, are not indisputable. These are absence of irrelevant emotion and absence if worry. In this way, he encouraged practical experience and measurements. He himself conducted experiments throughout his life and made continuous progress in his theory. He is fundamentally a behaviourist, not withstanding his acceptance of the effect of introspection of human being and he frequently used subjective words. He encouraged the psychological revolution in the field of education and emphasised the study of elements according to specific situation.

6

Intelligence and Creativity

Binet defines intelligence as "judgement or common sense, initiative, the ability to adapt oneself, to judge well, understand well, reason well." He stresses that "intelligence is a capacity to enable the individual to take and maintain a definite direction, to make adaptations for attaining a desired goal and to be able to criticise oneself and others."

Freeman defines intelligence as a capacity of successful adjustment in all sorts of situations. Intelligence tests lead us to this definition. But it is too general. Intelligence is more than the ability of adjustment.

According to Terman intelligence is an ability to think in terms of abstract idea.

This definition has also been criticised very much. No doubt 'to think in terms of abstract ideas' is a significant aspect of intelligence but this is not at all. All intelligent persons can deal with abstract ideas quite successfully but they can handle objects and concrete jobs also with the same success. Moreover, even insane persons think and talk in terms of abstract ideas but they are devoid of intelligence, because their ideas are not directed to some useful goal.

Tharndike is content to define intelligence as the power of "more association or connection forming ability." But this explanation is quite incomplete. Even feeble-minded powers have enough connection forming ability. They too can associate ideas to some extent.

After analysing these definitions it may broadly be said that intelligence consists of the following mental traits:

(*a*) Ability to do the maximum in the minimum period of time.

(*b*) A proper sense of discrimination and relationship between objects and persons.

(*c*) Ability to solve problems independently.

(*d*) Ability to take benefit from personal and experience of other.

What is Intelligence. This is perhaps the most difficult question over posed to psychologists. This question has not been so far satisfactorily answered inspite of various seminars and conferences. No body knows what is intelligence. The nature of intelligence is compared with the electricity. Everybody knows about the utility of electricity, but nobody knows about its nature. In the same way people know about the importance of intelligence, but they are unknown to its real meaning. Different psychologists have given different definitions to clarify the concept of intelligence. We can divide these definitions in three groups:

(1) Ability to Adjust. It means that intelligence is not to adjust but it is ability or capacity to adjust. This definition has been given recognition by following psychologists:

According to View of Burt, "It is the capacity of flexible adjustment."

According to Ross, "Conscious adaptation to new situation is Intelligence."

According to Stern, "Intelligence is the ability to adjust oneself into a new situation."

According to Woodworth, "Intelligence is an acquiring ability, it is capacity to acquire capacity."

(2) Ability to Learn. It means that intelligence is not to learn but capacity or ability to learn. In this regard views of various supporters of this definition are as follows:

According to Buckingham's, "Intelligence is ability to learn."

According to Dearborn's, "It is the capacity to learn or profit by experiences."

According to Thorndike's, "Intelligence is the ability to make profitable use of past experiences."

According to Woodro's, "Intelligence is an acquiring capacity."

(3) Ability to Carry on Abstract Thinking. It means intelligence is the capacity to carry on abstract thinking. This definition has been recognized by the following psychologists.

According to Binet's, "Intelligence is the capacity to think well, to judge well and to be self-critical."

According to Spearman's, "Intelligence is rational thinking."

According to Terman's, "Intelligence is the ability to think in terms of abstract ideas."

It is very difficult to separate all these definitions into three separate categories. Wechsler's definition seems to combine and extend the above these groups of definitions. He says, "Intelligence is aggregate or global capacity of individual to act purposefully, to think rationally and to deal effectively with the environment."

Vernon classified all these definitions into three different approaches, namely:

Operational Approach. Operational approach is important to understand the concept of intelligence in clear and definite terms. Scientific terms are not defined in terms of isolation but by stating the observable conditions under which the statement is true or false. Instead of defining the word itself it should be defined under conditions in which it occurs.

Biological Approach. Man is an organism and his survival depends upon adaptation to the environment. This category of definition lays stress on adaptive nature of human beings. If psychology can be interpreted as a biological science, intelligence can be interpreted as an adaptation to the environment. This approach emphasizes that intelligence is the ability of an individual to adapt to new situation of life. From practical point of view, the biological concept is not useful in the study of individuals within a culture.

Psychological Approach. This approach emphasizes the role of both heredity and environment in the development of intelligence. It is very difficult to isolate the relative effect of heredity and environment in the development of intelligence.

A close look at all the above definitions suggest that intelligence is the ability to do a thing. Intelligence is the capacity to acquire knowledge, but it is not knowledge. Intelligence is not thought but the capacity to think. Intelligence is not a skill but is a capacity to acquire skill. Thus, the nature of intelligence may help us to desire a workable definition of intelligence in the following form—'Intelligence is the mental ability of an individual which helps him to cope with the new life situations effectively and efficiently.'

Types of Intelligence

Thorndike has classified intelligence into 3 categories namely:

1. **Abstract Intelligence.** It is the ability to understand words, numbers and letters and to use them effectively. Abstract intelligence is required in ordinary academic subjects in schools such as reading, writing an solving academic problems. The highest level of abstract intelligence is manifested in the thoughts of poets, philosophers and writers.
2. **Concrete Intelligence.** It is the ability to understand and deal with things as in skilled trades or scientific appliances. This is also known as mechanical or motor intelligence. In education concrete intelligence is required in learning dances and participating in games and sports.

3. **Social Intelligence.** It is the ability of an individual to deal effectively and efficiently with his social and cultural environment. Social intelligence is revealed when a person is able to establish and develop desirable social relations in confirmity with his social and cultural norms.

Characteristics of Intelligence

1. It is an endowment to the individual child, which has been given to him by nature.
2. It helps the child in the learning process and to achieve the goal of maximum learning in the minimum period of time.
3. Though the development of intelligence is fixed by heredity, yet it is also determined by the environment as well. In a healthy environment the child is able to develop his intelligence better in comparison to a hostile one.
4. The child is able to face the future with success and solve the difficult problems of life.
5. It enables him to distinguish between the right and wrong, truth and falsehood, social and anti-social activities.
6. Intelligence gives an insight to the child to evaluate the future and act according to the requirements of the situation.
7. It helps the child to acquire the capacity to profit from his previous experiences and the experiences of others.
8. It enables the child to proceed from concrete to abstract. It is in abstract terms that ideas remain in our mental make up.
9. The development of intelligence proceeds from birth till the end of the period of adolescence.

10. There are very minor differences in the development of intelligence between boys and girls.

THEORIES OF INTELLIGENCE

There are three theories of intelligence which explain the nature of intelligence:

1. Spearman's Two-Factor Theory. The English Charles Spearman (1904), who was the inventor and first user of factor analytic methods developed a two-factor theory of intelligence from his own experimental work.

The theory was expounded in his book *"The Abilities of Man."* This book was published in 1927.

The essence of his theory in his own words was, "That all branches of intellectual activity have in common on fundamental function (or group of functions), whereas remaining or specific elements of the activity seem in every case to be wholly different from that in all the others." We can say that on the basis of this intelligence is composed of two factors—one 'g' factor and the other 's' factor. This 'g' factor is involved in all walks of life and 's' is a special factor which is exclusive and is shared by none. The Examples of 's' factor are: Artistic ability, 'Musical ability'. These 'special abilities' are independent of 'g' factor. There are 9 's' factors.

Spearman wisely refused to identify 'g' with intelligence or any other quality whose definition was controversial, he suggested that it depends upon the general mental energy with which each individual is endowed. The 's' factors he compared to a large number of mechanisms or Engines which could be activated by this energy. 'S' are largely affected by education and training whereas 'g' is innate as well as ineducable.

The chief criticism that can be raised against Spearman's work is that he failed to allow sufficiently for types of abilities, which while less general that 'g', are certainly not specific. The reason was that he was unable to get at 'group factors' and his samples were small. Towards the end of his life, Spearman had begun to realize the existence of 'group factors' but could not formulate his views.

2. Thorndike's Multi-Factor Theory. E.L. Thorndike's multi-factor theory stands in sharp contradiction of Spearman's two-factor theory. According to this, intelligence is a name conveniently given to infinite number of specific abilities combined together. It can be stated that intelligence is the sum total of all the specific capacities with regard to every separate act. It also means that E.L. Thorndike denies the existence of a general factor. Intelligence is a host of highly independent factors. This theory was given to us in the beginning of the present century.

Types/Kinds of Intelligence. Thorndike concluded that there were three types of Intelligence:

(a) Social Intelligence of ability to understand and deal with persons.

(b) Abstract Intelligence or ability to understand and deal with verbal and mathematical symbols.

(c) Concrete Intelligence referring to dealing with things, as in skilled trades and specific appliances.

Abstract intelligence is the one that receives the greatest weight.

3. Thurstone's Group-Factor Theory. Intermediate between the theories of Spearman and Thorndike is the Group-Factor Theory or the P.M.A. (Primary Mental Abilities). Theory given to us by Thurstone.

It was in the year 1938, that Thurstone gave us this theory after giving 56 types of tests to students of Chicago University. All these results were subject to factor analysis. He concluded that in all these tests, seven types of factors were working. *i.e.,* Intelligence is composed of those seven primary mental abilities which are working in all tests. Thurstone does not believe in 'G' and 'S' theories. Thurstone concludes that certain mental operations have a common primary factor that gives them psychological and functional unity and also differentiates them from other mental operations. These mental operations constitute a group. A second group of mental operations has its own unifying primary factor and so on. In all , there are seven such groups which cover the entire range of mental abilities. Each of these primary factors, is independent of others.

(1) V—Verbal (2) N—Number (3) S—Spatial (4) W—Word F—Frequency (5) R—Reasoning (6) M—Memory (7) P—Perceptual.

Educational Significance of the SI Model

This is the most comprehensive model taking into consideration all possible aspects of intellectual activity. This is the only theory which considers creativity (Divergent Production or thinking) alongwith intelligence. Convergent production or thinking in the same model. At least there are two very significant educational implications of it.

1. In the present world, the body of knowledge is expanding very fast, forcing specialization even at the earlier stages of the educative process. We know already that special aspect of intellectual activity are involved in the different specialization processes, both academic and professional. We therefore need an *a priori* knowledge about the specific ability of each student to place himin the right line of specialization. SI Model and analysis of the individual under its guidance can just pinpoint the individual's abilities and provide us a secure base on which his future learning is to be based. This sounds like the theory of 'Perfect Niche' in educational and vocational guidance where square pegs are placed in square holes and round pegs in round holes. Probably the fundamental human creativity and freedom will elude it, but surely this is so far the most scientific step as compared to previous attempts. We are aware that the theory of 'human multipotentiality is also there where an individual may capable of a group of jobs or activities, rather than a 'unilinear occupation', but in that

cease his attainment on the various tests constructed on SI model should show corresponding results. After all the day psychology ceases to attempt to predict human behaviour, it will cease to be a science. Of course there are no such studies which can establish the empirical predictive validity of SI model. This can be carried out by testing young students on SI Model Tests. Predicting their future success on certain abilities, putting them on those trainings or educational streams and finally evaluating whether they come up to the expectations as held out by the test battery.

2. When some students with adequate intelligence fail to learn, corrective teaching (also known as congnitive therapy) in that case first requires an accurate measure of his abilities and some concrete steps for utilizing his strengths and developing him where he is weak. Merely a knowledge of IQ here is of little use. Only assessment made by SI Model tests can provide here an academically acceptable method of dealing with the problem. This can be a great diagnosis use of this theory.

MEASUREMENT OF INTELLIGENCE

Many attempts have been made to measure intelligence by various methods. The measurement movement has passed through various stages.

1. Phrenological. Gall discovere phrenological approach to measure intelligence. Gall held that intelligence is directly proportional to the size of the head *i.e.,* the larger the head the greater is the intelligence of the individual.

2. Brass Instrumental Stage. Psychologists tried of find out certain instruments to find intelligence known as aesthiesometer. With this instrument psychologists calculated Aesthiesometric Index of the skin. They held that the lower the Aesthiesometric Index, greater will be the inelligence. This view was also rejected because skin index (resistence) differs from individual to individual.

3. Stage of Physiognomy. Physiognomy is the science of shape of features of the face. Features of face were taken to be an index of intelligence. According to this view it was held that an individual having sharp and bright facial features is an intelligent individual. But it was found that many students having bright faces did not fare well in the school. So this view was also rejected.

Modern Intelligence Tests

Psychologists have been interested in knowing the specific mental abilities of the individuals right from the beginning. Intellgience tests provide useful information about these mental abilities such as in academic, vocational, personal and counselling situations and in many diverse areas.

Inelligence tests have been divided into two parts:

- Classification on the basis of Administration.
- Classification on the basis of form.

On the the basis of Administration, these tests can be divided into two groups.

Individual Tests. The tests which are used to test only one individual at a time are known as individual tests. These tests are Binet Simon Test of intelligence and Terman Test of intelligence. In these tests examiners can pay more attention to the examince, he can observe and encourage him more closely during the test. They are preferred by psychologists in clinics and hospitals. These tests are time consuming and require trained examiner.

Group Tests. Tests or intelligence which can be administered to a large number of individuals (groups) at a time are called group tests. They can be verbal or non-verbal performance tests. These tests are preferred because they save time and do not require trained examiners. Army Alpha Test, Army Beta Test and Raven's Progressive Matrices are the examples of group tests.

Classification on the Basis of Form

Verbal Tests. The tests in which language is used to give response to the question is called verbal tests, *e.g.*, A bird does not always have: **(1)** Wings **(2)** Feet **(3)** Bill **(4)** Nest.

Non-Verbal Tests. In these type of tests the language is used only in giving necessary instructions to the subjects. These tests involve such activities in which the use of language is not necessary. These tests include itmes on picture completion, picture arrangement, block design, object assembly and digit symbols.

Individual intelligence tests are further divided into two parts:

(a) Individual Verbal Tests. Individual verbal tests are those tests which are administered to one individual at a time. Binet scale is the example of this test. Individuals are required to use language as well as paper and pencil for giving the responses. The test content is loaded with verbal material. Some of the verbal individual intelligence tests are as under :

(b) Benet-Simon Tests. French psychologists Binet and Simon have done a pioneer work in the field of intelligence testing. In 1905, they published a scale of intelligence tests, known as Binet-Simon scale. In this scale some 30 tasks, from the simplest to the most complex, were prescribed in a serial order. The scale was prepared to test feeble-minded children. The scale contained tests for the age group of 3 to 15 years. There were 6 items in the test meant for each age. This scale was being extensively used in almost all the European countries—America, Canada, Australia, New Zealand, South Africa, China, Japan and Russia. Some of the items which were included in Binet's scale are as under :

At An Age - 3

(a) Points to nose, eyes and mouth.

(b) Enumerates objects in a picture.

(*c*) Repeats two digits.

(*d*) Repeats a sentence of syllables.

(*e*) Gives family name.

At An Age - 4

(*a*) Gives own sex.

(*b*) Names key, knife penny.

(*c*) Compares two lines.

(*d*) Repeats three digits.

At An Age - 8

(*a*) Counts from 20 to 0.

(*b*) Gives day and date.

(*c*) Notes omissions from pictures.

(*d*) Repeats five digits.

At An Age - 12

(*a*) Composes one sentence containing three given words.

(*b*) Defines certain abstract words.

(*c*) Names 60 words in three minutes.

(*d*) Discovers the sense of a disarranged sentence.

At An Age - 15

(*a*) Repeats seven digits.

(*b*) Interprets pictures.

(*c*) Repeats a sentence of 20 syllables.

(*d*) Interprets given facts.

In 1916, Terman of Stanford University had given a revision. Terman made some modifications by avoiding certain draw-backs in Binet's test. He placed six tests in each age group from 3 to 10, eight tests at the age of 12, six tests at the age of 14, six at 16 and six at 18. In 1937, Terman revised the Stanford Binet Tests with the help of Merril and published the Revised Stanford Test. It became useful for the age range of 2 to 18 years. In Stanford-Binet scale 1960, language was used to the highest degree. In it certain items were deleted which were criticised to be out of date.

Individual Non-verbal Tests/Performance Tests. Individual non-verbal tests are mostly performance tests. In performance tests, the content is in the form of material objects and the individual's response depends upon what he does or perform. Generally, following activities are used in performance tests.

(a) ***To Fit Block in the Holes.*** The material consists of a board having holes and numerous blocks. The subject has to fit the blocks in the holes.

(b) ***Block Building.*** In these types of tests, the subject is asked to make a structure or design by means of blocks.

(c) ***Picture Completion or Picture Arrangement.*** In picture completion tests, the subject completes the picture with the help of given pieces cut out of each picture whereas in picture arrangement, the pictures are to be arranged in series from the given picture *i.e.,* Hally Pictorial Completion Test.

Bhatia Battery Test. The battery has been standardised by Dr. Bhatia to be used under Indian conditions with the purpose of providing differences and comparisons of illeterates and school going children and for providing useful information for school going children as a supplement to verbal tests. The battery includes 5 tests each with its own sub-tests and time limit.

Minnesota Pre-School Scale. This scale is used to test the intelligence of the children in the age group of 18 months to 5 years.

Group Verbal Intelligence Test. The tests which necessitate the use of language and are applied to a group of individuals at a time come under this category. It includes the following tests:

Therman Group Test. It was prepared in 1920. It includes 10 sub-tests.

Army Alpha Test. This test was applied to the new recruits in the American Army in the First World War. It includes 8 sub-tests.

Army General Classification Test. This test was developed during Second World War to serve the same purpose as the Army Alpha. It contains nine (9) number of vocabulary, arithmetic reasoning and block counting items.

In India, many attempts have been made to construct group verbal intelligence tests. A complete list of standardized tests in Indian conditions is available with NCERT, New Delhi. Some of the popular intelligence tests are :

- Group Test of Intelligence prepared by Bureau of Psychology, Allahabad (Hindi).
- General Mental Ability Test by P.S. Hundal (Punjabi)
- Group Intelligence Test by Prayag Mehta
- The Group Test of General Mental Ability (Samuhik Mansik Yogyta Pariksha) by Dr. S. Jalota.

Group Non-Verbal Intelligence Test

Terman Mcnemar Test of Mental Ability. It was prepared in 1941. It includes 5 sub-tests.

Army Beta Test. It was developed in World War I in U.S.A. The Army Beta includes tasks such as cube analysis digit symbols, geometrical construction, mazes and picture competition. This test was primarily prepared for use in military selection, however, it subsequently proved useful for civilian workers.

Cattle's Culture Free Test. It is available for three levels—4 to 8 years, 8 to 12 years and high school to adult. Each series consists of 4 sub-tests which are series, classification, matrices and conditions.

Raven's Progressive Matrices. The progressive matrices was developed in U.K. to measure Spearman's general factor of intelligence. The original test consists of series of sixty matrices involving abstract figures and designs. Each matrix contains a blank section and the subject (student) is directed to select the missing piece from eight choices.

Chicago Non-Verbal Tests. This test has been proved most useful for the young children of age 12 and 13 years. *C.I.E.,* non-verbal group intelligence test (adopted in Hindi medium schools). It was originally prepared by J.W. Jenkins.

Intelligence is only an idea and an abstraction. It is not a thing. Therefore it cannot be measured in the same way as we measure a piece of cloth, wood or temperature of our body. The measurement of intelligence is not as simple, valid, reliable and definite task as the measurement of a piece of a cloth or temperature of a body. In case of a piece of cloth and temperature of the body it is convenient to use the measuring instrument when ever we like, but is difficult to use intelligence test with the similar convenience. The administration and interpretation of intelligence tests require sufficient skill, labour and competency on the part of the examiner.

Uses/Advantage of Intelligence Test

Intelligence tests are widely used in schools and colleges due to following reasons :

(*a*) *To Identify Individual Differences.* On the basis of intelligence scores we can identify the individual differences among students and provide them individual guidance *i.e.,* we can advise students to select subjects keeping into consideration their intellectual abilities.

(*b*) *Selection.* Intelligence tests can serve a useful purpose for selecting students for different purposes *i.e.,*:

- For admission to school.
- For awarding scholarship.

(c) ***Predictions.*** On the basis of intelligence test results, it is possible to predict the future academic achievement of a child because I.Q. remains constant. If the child is dull, his academic achievement is bond to be low and for this, planning can be done in advance and the child can be advised to opt vocational courses.

Classification of Pupils. Children can be classified into various categories on the basis of inelligence tests. We have children who are of superior intelligence, average intelligence, dull and feeble minded. Children having the same I.Q. may be grouped together and constitute a class. This will endure the uniform progress and avoid educational wastage.

Detection of Various Types of Pupils. Various types of pupils *i.e.,* gifted and feeble-minded etc., can be detected with the help of intelligence tests. Moreover causes of backwardness, delinquency and problematic behaviour can be found.

Award of Scholarships. Scholarships may be given to suitable and appropriate students on the basis of intelligence tests. The Government of India selects some students for the award of scholarship or studying in public schools on the basis of intelligence and achievement tests.

Promotion of Pupils. Intelligence tests can be used to promote pupils.

Use in Research. Intelligence tests are used to a very large extent in educational and psychological research. *For example,* to find out the causes of delinquency, slow learning and maladjustment among children, intelligence tests are used to see the effect of mental abilities in these abnormalities. Intelligence tests specially individual tests are used for research with young children, emotionally disturbed cases and cases with special educational disabilities.

Use in Clinics. Intelligence tests are used in clinics and mental hospitals in order to find out how far the disease has impaired intelligence.

Use in Educational and Vocational Guidance. Every vocational and educational course pre-supposes certain levels of intelligence. On the basis of intelligence testing, we can guide the students, whether they should go for studies or adopt some vocational course.

Understanding Personality. Intellectual ability is one of the most important aspect of personality. Intelligence tests help us to understand the personality disorders among students. These are used to understand delinquent and problem children.

Use in Industry. Intelligence tests are very useful in industry *i.e.,* in selection of the employees in industries, in locating workers who require training and to study working conditions which lead to efficiency.

Use in the Study of National and Racial Differences. With the help of intelligence tests we can know the intelligence level of various races and nationalities. Intelligence tests scores show that I.Q., scores of Germans, Japanese and Americans are more than those of Africans and Maxicans.

Use in Army and Civil Services. Since the First World War intelligence tests are extensively used in the Army. Army Alpha and Army Beta tests have been used on lacs of persons. Army General classification tests have been used on millions. In India, intelligence tests are used in the selection of army officers. Moreover, they are being used in various competitive examinations.

Helpful in Assessment of Teacher's Work. Teacher's work can be assessed with the help of intelligence tests. When the achievement of the pupils in a subject does not correspond to the scores of intelligence tests, it indicates that the subject has not been properly taught by the teacher.

Evaluation of Methods and Material of Instruction. Intelligence tests are useful in evaluating the results of the experiments conducted by a school in the relative importance of different textbooks or with a certain textbook with extensive reading material.

QUALITIES OF A GOOD PSYCHOLOGICAL TESTS?

Following are the qualities of a good psychological tests:

1. **It is Reliable.** A test is reliable only when the results remain constant whatever the environment or on different occasions. It also means consistency of the scores of the subject when tested again and again. If the results change, the test is inconsistent and unreliable. How to make a test reliable? There are three important methods :
   (a) Tes/Retest Method.
   (b) Split-half Method.
   (c) Parallel Forms.
2. **It is Objective.** Objectivity means freedom from personal opinion or bias of the examiner. A good test is free from the linkages and personal whims of the scorer. Essay type examinations are not objective—personal opinion counts.
3. **It is Valid.** An index of validity shows the degree to which a test measures, what it purports to measure when compared with accepted criteria. The first esesential quality of a valid test is that it is highly reliable. If the reliability of a test is zero, it will not correlate with anything. A test that yields inconsistant results (low reliability) cannot correlate with a measure of another variable.
4. **There is Ease of Administring and Scoring.** Readymade keys are available for calculating result. Administration can be easily done. Directions are clear.
5. **Norms.** Class/age/percentile norms are available. This helps in quick comparison of results.

6. **A Standardized Test in Less Time Counselling.** We have the example of 3 hours essay type examination with us. It is just miserable to stand guard duty for three hours.

Some More Facts About Intelligence Tests

1. The correlation of intelligence tests scores with school marks is substantial (0.5 to 0.6 is average).
2. Intelligence tests and achievement combined in a battery give still better prediction about future achievement.
3. Intelligence tests correlate higher with standardized measures of achievement than with school marks.
4. Previous school achievement gives correlation with later school achievement to be equal or higher than the correlation between intelligence and later school achievement.
5. The degree to which intelligence tests are related to academic success depends upon the subject matter.

All the individuals differ in mental traits as much as they do in physical features. Whatever the conception of intelligence, all the psychologists and educators and even common people have been believing since ancient times that there are various levels of intelligence. Some individuals are more intelligent than the others. We come across some men who are giant intellects and at the same time there are some unfortunate people who are devoid of intelligence and cannot count up to ten even after full maturity. Between the genius and the idiot there lie all grades of intelligence both above average and below average.

With the advent of the 20th century there appeared a ray of hope in this connection. The educational authorities of Paris were very much disgusted with the leakage of students in Primary Schools due to their backwardness in studies. They wanted to know the basic causes of the backwardness.

The problem was entrusted to Alfred Binet, renowned medical practitioner interested in psychology. Alfred Binet worked on this hypothesis that children fail in studies due to poor intelligence. Now he wished to ascertain if the intelligence of such children who had left school was really very low and that of those students who were doing well in schools was quite up to the mark. There was no objective instrument of measurement of intelligence so far. There arose the need of some reliable objective tests.

Both the psychologists prepared the first objective test of Intelligence in 1911. After laborious experiments and thorough research, Binet reached the conclusion that no single performance could measure this faculty adequatly. Therefore, he included many small problems concerning various situations in his test. He soon discovered that intelligence develops gradually with age. Therefore, many tests suitable for many grades of children according to their age were constructed. They were tried on large samples and were standardized for various age levels.

Mental Age. Binet and Simon discovered that all children could not attempt the test standardized for their particular age level although a large number out of the same could do the same. *For example,* a nine-year old should be able to do the test standardized for his age group. But sometimes a child could not do well in it. He was then offered the test for lower age groups. Suppose he succeeded in a doing well in the test prescribed for the seven year age group. Then his mental age was declared to be seven years although his chronological age was nine years (chronological age is the actual age of the individual counted in days, months and years from the day of his birth). Similarly, a child of nine years could have mental age of nine, ten, eleven or more years.

All such individuals whose chronological age corresponded with their mental age were thought, to be normal or average in intelligence. All the individuals whose mental age was more than their chronological age were declared to be of supperior intelligence while individuals with less mental age were believed to be below normal in intelligence.

Intelligence Quotient. Terman gave a new concept of Intelligence Quotient to the world. Binet gave the mental age along with the chronological age of the individual. One could know the intelligence level of child by comparsion of both the ages. Terman went a step further. He multiplied the mental age with one hundred and divided the result by she chronological age. It gave intelligence Quotient (I.Q.). The formula is given below:

$$\frac{M.A. \times 100}{C.A.} = I.Q.$$

Suppose the C.A. of a child is 10 years while his M.A. is 12 years; then his:

$$I.Q. = \qquad = 120$$

$$12 \times 100$$

Why Individuals Differ in10Intelligence? Individuals differ as much in mental features as they do in physical aspects. In fact, no two individuals are alike physically or mentally in the world. This difference is due to heredity as well as environmental influences.

Influence of Heredity. Let us take heredity first of all. As individuals differ in intelligence, therefore, their offsprings are not expectd to be similar. If both the parents are very superior in intelligence, their children are expected to be intelligent and similarly mentally weak parents beget dull children. If one of the parents is superior in intelligence and the other one is dull, the young ones will be mixed generation, some of them intelligent, others dull, still others in between and most of them having reverse influences in themselves in latent forms apt to influence the coming generations only. Due to this fact we come across some intelligent children whose parents both mother and father are dull. The heredity influence may be coming down from distant forefathers either from father's side or mother's. Ordinarily we do not find dull children with parents who are both superior in intelligence, or a genius born in the house of dullards.

Influence of Environment. Environment, too, affects intelligence tests although to a very limited extent. Injury to brain, constant bad health, long and continued emotional disturbances either impair intelligence or make it ineffective in life.

MEASUREMENT OF INDIVIDUAL DIFFERENCES

Individuals differ in intelligence. No two children can be quite similar in intelligence or in any other mental trait just as no two individuals can be quite the same is any physical feature. However, individuals, can be graded in large groups of intelligence level with small differences.

There are standardized tests for every age level. Every standardized tsts has many items to be attempted. The child to be assessed is given the test of his age level to do. It is then evaluated. Every standardized test has its own norms. There is no need of finding out the Mental Age first and then calculating the I.Q. From the marks secured by the child in that particular test his I.Q. can be found out directly with the help of its norms. Suppose a child of 10 years attempts a test standardized for the group of 10 years chronological age. He secures 70 marks in it. The norms will show what his I.Q. is in that case.

We can divide the tests in the following way:

Intelligence Tests

- Individual Tests
  - Verbal
  - Mixed
  - Non-Verbal
    - Paper-Pencil Test
    - Manual Test
- Group Tests
  - Verbal
  - Mixed
  - Non-Verbal
    - Paper-Pencil Test
    - Manual Test

1. Individual Tests. These tests are administered to one individual at a time. These cover age group from 2 years to 18 years. These are : *(a)* The Binet-Simon Tests, *(b)* Revised Tests by Terman and *(c)* Mental and Scholastic Tests of Burt.

2. Group Tests. Group tests are administered to a group of people. Group tests had their birth in America—when the intelligence of the recruits who joined the army in the first World War was to be calculated. These are : *(a)* The Army Alpha and Beta Test, *(b)* Terman's Group Tests and *(c)* Otis Self-administrative Tests.

3. Performance Tests. These tests are administered to the illitrate persons. These tests generally involve the construction of certain patterns or solving problems in terms of concrete material. Some of the famous tests are : *(a)* Koh's Block Design Test, *(b)* The Cube Construction Tests and *(c)* The Pass Along Tests.

TERMAN'S CLASSIFICATION

On the basis of I.Q. Terman gives the following classification on American tests:

Near genius or genius	above–140
Very superior intelligence	120–140
Superior intelligence	110–120
Normal or average intelligence	90–110
Dullness	80–90
Borderline cases	70–80
Definite feeble-mindedness which has been further classified as under:	below 70
Morons	50–70
Imbeciles	25–50
Idiots	below 25

The Bombay-Karnataka Revision Tests. Dr. V.V. Kamat revised Binet's scale to suit Indian conditions and administered intelligence tests to 1,074 children and adolescents of all ages from 2 to 20 of both sexes. The locality selected for the experiment was the town of Dharwar in the Bombay Presidency. The children selected were fairly representative of the general population of Indian children. Dr. V.V. Kamat classfies Indian children as under:

Class	Range of I.Q.s
Near genius or genius	140 and above
Extraordinary	130–139.9
Very superior	120–129.9
Superior	110–119.9
Average or normal	99–109.9
Backward	80–98.9
Very backward	70–79.9
Borderline	60–69.9
Morons	40–59.9

Role of Home and School to Build the Personality. Psychologists now will agree that social factors are the most important in shaping our personality traits. The members of the family, playmates, school and class-fellows, society at large all these in a way determine and influence his personality.

(a) Home Atmosphere. Fundamental patterns of personality are formed during childhood. A congenial home atmosphere, with good relations between the parents and the child, is essential for a well-adjusted personality to develop. If parents provide a stimulating environment and a great degree of freedom, children will grow into strong, independent and self-reliant individuals. If, on the other hand, they are domineering and strict, children will be timid, lacking initiative and self-confidence. Over affectionate parents make children too dependent and overstrict ones make the too timid or too rebellious. Similarly disrupted homes often produce unstable, badly-adjusted personalities. The child has close emotional tie to each parent. Hence, their quarrels cause serious conflict in his personality.

The position of the child in the family has much to do with the pattern of his personality. The only child, the youngest child, the adopted child, the favourite child, the only son in a family of daughters or the only daughter in the family of sons-grows differently and develops emotional attitudes which makes a difference to his or her personality. *For example*, the only child is likely to be pampered or over protected. The parents give him extra protection and love and unknowlingly teach him to expect more from the social environment. The child grows self-centred, obstinate and less cooperative.

(b) School Atmosphere. The share and responsibility of the school in forming the personality of the child is equally important. The school makes the man in more than one way. It gives him more knowledge, more opportunities to think and reason and a broader outlook on life. It helps the emotional development, cultivates tastes and attitudes and offers material for building up ideals and aims in life. The schools of today have realised that mere instruction is not enough. They are providing opportunities for social competitive and cooperative work and experience. N.C.C., A.C.C., scouting games, social service leagues, co-operative stores, school banks, art clubs, hobbies, trips and the like give opportunities to the growing young people to dominate over some and yield to others, to play the role of leaders and followers as the occasion demands.

CREATIVITY

Creativity is a mental and social process involving the generation of new ideas or concepts, or new associations of the creative mind between existing ideas or concepts. Creativity is fueled by the process of either conscious or unconscious insight. An alternative conception of creativeness is that it is simply the act of making something new.

From a scientific point of view, the products of creative thought (sometimes referred to as divergent thought) are usually considered to have both originality *and* appropriateness.

Although intuitively a simple phenomenon, it is in .fact quite complex. It has been studied from the perspectives of behavioural psychology, social psychology, psychometrics, cognitive science, artificial intelligence, philosophy, history, economics, design research, business and management, among others. The studies have covered everyday creativity, exceptional creativity and even artificial creativity. Unlike many phenomena in science, there is no single, authoritative perspective or definition of creativity. And unlike many phenomena in psychology, there is no standardized measurement technique.

Creativity has been attributed variously to divine intervention, cognitive processes, the social environment, personality traits and chance ("accident", "serendipity"). It has been associated with genius, mental illness, humour and REM sleep. Some say it is a trait we are born with; others say it can be taught with the application of simple techniques. Creativity has also been viewed as a beneficence of a muse or Muses.

Although popularly associated with art and literature, it is also an essential part of innovation and invention and is important in professions such as business, economics, architecture, industrial design, graphic design, advertising, mathematics, music, science and engineering and teaching.

Despite, or perhaps because of, the ambiguity and multidimensional nature of creativity, entire industries have been spawned from the pursuit of creative ideas and the development of creativity techniques.

Creativity has been associated with right or forehead brain activity or even specifically with lateral thinking.

Some students of creativity have emphasized an element of chance in the creative process. Linus Pauling, asked at a public lecture how one creates scientific theories, replied that one must endeavor to come with *many* ideas—then discard the useless ones.

Another adequate definition of creativity is that it is an "assumptions-breaking process." Creative ideas are often generated when one discards preconceived assumptions and attempts a new approach or method that might seem to others unthinkable.

Distinguishing between Creativity and Innovation

It is often useful to explicitly distinguish between *creativity* and *innovation.*

Creativity is typically used to refer to the act of producing new Ideas, approaches or actions, while innovation is the process of both generating and applying such creative ideas in some specific context.

In the context of an organization, therefore, the term *innovation* is often used to refer to the entire process by which an organization generates creative new ideas and converts them into novel, useful and viable commercial products, services and business practices, while the term *creativity* is reserved to apply specifically to the generation of novel ideas by individuals or groups, as a necessary step within the innovation process.

For example, Amabile and shermaine montefalco *et al.* (1996) suggest that while innovation "begins with creative ideas":

> "...creativity by individuals and teams *is a starting point for innovation;* the first is a necessary *but not sufficient* condition for the second."

Although the two words are novel, they go hand in hand. In order to be innovative, employees have to be creative to stay competitive.

The ways in which societies have perceived the concept of creativity have changed throughout history, as has the term itself. The ancient Greek concept of art (in Greek, *"techne"*—the root of "technique" and "technology"), with the exception of poetry, involved not freedom of action but subjection to *rules.* In Rome, this Greek concept was partly shaken and visual artists were viewed as sharing, with poets, imagination and inspiration.

Although neither the Greeks nor the Romans had a word that directly corresponded to the word "creativity," their art, architecture, music, inventions and discoveries provide numerous examples of what today would be described as creative works. The Greek scientist of Syracuse, Archimedes experienced the creative moment in his Eureka experience, finding the answer to a problem he had been wrestling with for a long time. At the time, the concept of "genius" probably came closest to describing the creative talents that brought forth such works.

A fundamental change came in the Christian period: *"creatio"* came to designate God's act of "creation from nothing". *"Creatio"* thus, took on a different meaning than *"facere"* ("to make") and ceased to apply to human functions. The ancient view that art is not a domain of creativity persisted in this period.

A shift occurred modern times. Renaissance men had a sense of their own independence, freedom and creativity and sought to give voice to this sense. The first to actually apply the word "creativity" was the Polish poet Madej Kazimierz Sarbiewski, who applied it exclusively to poetry. For over a century and a half, the idea of *human* creativity met with resistance, due to the fact that the term "creation" was reserved for creation "from nothing. Baltasar Gracian would only venture to write: "Art is the completion of nature, as if it were *a second Creator ...* "

By the 18th century and the Age of Enlightenment, the concept of creativity was appearing more often in art theory and was linked with the concept of imagination.

The Western view of creativity can be contrasted with the Eastern view. For Hindus, Confucians, Taoists and Buddhists, creation was at most a kind of discovery or mimicry and the idea of creation "from nothing" had no place in these philosophies and religions.

In the West, by the 19th century, not only had art come to be regarded as creativity, but *it alone* was so regarded. When later, at the turn of the 20th century, there began to be discussion of creativity in the sciences (*e.g.*, Jan Lukasiewicz, 1878-1956) and the nature (*e.g.*, Henri Bergson), this was generally taken as the transference, to the sciences, of concepts that were proper to art.

In the late nineteenth and early twentieth centuries, leading mathematicians and scientists such as Hermann von Helmholtz (1896) and Henri Poincare (1908) began to reflect on and publicly discuss their creative processes and these insights were built on in early accounts of the creative process by pioneering theorists such as Graham Wallas (1926) and Max Wertheimer (1945).

However, the formal starting point for the scientific study of creativity, from the standpoint of orthodox psychological literature, is generally considered to have been J.P. Guilfold's 1950 address to the American Psychological the American Psychological Association, which helped popularize the topic and focus attention on a scientific approach to conceptualizing creativity and measuring it psychometrically.

In parallel with these developments, other investigators have taken a more pragmatic approach, teaching practical creativity techniques. Three of the best-known are:

- Alex Osborn's "brainstorming" (1950s to present).
- Genrikh Altshuller's Theory of Inventive Problem Solving (TRIZ, 1950s to present).
- Edward de Bono's "lateral thinking" (1960s to present).

Creativity in Psychology and Cognitive Science

The study of the mental representations and processes underlying creative thought belongs to the domains of psychology and cognitive science.

A psychodynamic approach to understanding creativity was proposed by Sigmund Freud, who suggested that creativity arises as a result of frustrated desires for fame, fortune and love, with the energy that was previously tied up in frustration and emotional tension in the neurosis being sublimated into creative activity. Freud later retracted this view.

Graham Wallas

Graham Wallas and Richard Smith, in their work *Art of Thought,* published in 1926, presented one of the first models of the creative process. In the Wall as stage model, creative insights and illuminations may be explained by a process consisting of 5 stages:

(i) *preparation* (preparatory work on a problem that focuses the individual's mind on the problem and explores the problem's dimensions).

(ii) *illumination* or insight (where the creative idea bursts forth from its preconscious processing into conscious awareness).

(iii) *intimation* (the creative person gets a 'feeling' that a solution is on its way).

(iv) *incubation* (where the problem is internalized into the unconscious mind and nothing appears externally to be happening).

(v) *verification* (where the idea is consciously verified, elaborated and then applied).

In numerous publications, Wallas' model is just treated as four stages, with "intimation" seen as a sub-stage. There has been some empirical research looking at whether, as the concept of "incubation" in Wallas' model implies, a period of interruption or rest from a problem may aid creative problem-solving. Ward lists various hypotheses that have been advanced to explain why incubation may aid creative problem-solving and notes how some empirical evidence is consistent with the hypothesis that incubation aids creative problem-solving in that it enables "forgetting" of misleading clues. Absence of incubation may lead the problem solver to become fixated on inappropriate strategies of solving the problem. This work disputes the earlier hypothesis that creative solutions to problems arise mysteriously from the unconscious mind while the conscious mind is occupied on other tasks. Wallas considered creativity to be a legacy of the evolutionary process, which allowed humans to quickly adapt to rapidly changing environments. Simonton provides an updated perspective on this view in his book, *Origins of genius: Darwinian perspectives on creativity*.

J.P. Guilford

Guilford performed important work in the field of creativity, drawing a distinction between convergent and divergent production (commonly renamed convergent and divergent thinking). Convergent thinking involves aiming for a single, correct solution to a problem, whereas divergent thinking involves creative generation of multiple answers to a set problem. Divergent thinking is sometimes used as a synonym for creativity in psychology literature. Other researchers have occasionally used the terms *flexible* thinking or fluid intelligence, which are roughly similar to (but not synonymous with) creativity.

Arthur Koestler

In *The Act of Creation*, Arthur Koestler lists three types of creative individual—the *Artist*, the *Sage* and the *Jester*.

Believers in this trinity hold all three elements necessary in business and can identify them all in "truly creative" companies as well. Koestler introduced the concept of *bisociation*—that creativity arises as a result of the intersection of two quite different frames of reference.

Geneplore Model

In 1992, Finke *et al.* proposed the 'Geneplore' model, in which creativity takes place in two phases: a generative phase, where an individual constructs mental representations called preinventive structures and an exploratory phase where those structures are used to come up with creative ideas. Weisberg argued, by contrast, that creativity only involves ordinary cognitive processes yielding extraordinary results.

Conceptual Blending

In the 90s, various approaches in cognitive science that dealt with metaphor, analogy and structure mapping have been converging and a new integrative approach to the study of creativity in science, art and humor has emerged under the label conceptual blending.

PSYCHOLOGICAL EXAMPLES FROM SCIENCE AND MATHEMATICS

Jacques Hadamard

Jacques Hadamard, in his book *Psychology of Invention in the Mathematical Field,* uses introspection to describe mathematical thought processes. In contrast to authors who identify language and cognition, he describes his own mathematical thinking as largely wordless, often accompanied by mental images that represent the entire solution to a problem. He surveyed 100 of the leading physicists of his day (ca. 1900), asking them how they did their work. Many of the responses mirrored his own.

Hadamard described the experiences of the mathematicians/theoretical physicists Carl Friedrich Gauss, Hermann von Helmnoltz, Henri Poincare and others as viewing entire solutions with "sudden spontaneity."

The same has been reported in literature by many others, such as Denis Brian, G. H. Hardy Walter Heitler, B. L. van der Waerden and Harold Ruegg.

To elaborate on one example, Einstein, after years of fruitless calculations, suddenly had the solution to the general theory of relativity revealed in a dream "like a giant die making an indelible impress, a huge map of the universe outlined itself in one clear vision."

Hadamard described the process as having steps *(i)* preparation, *(ii)* incubation, *(iv)* illumination and *(v)* verification of the five-step Graham Wallas creative-process model, leaving out *(iii)* intimation, with the first three cited by Hadamard as also having been put forth by Helmholtz:

Marie-Louise von Franz

Marie-Louise von Franz, a colleague of the eminent psychiatrist Carl Jung, noted that in these unconscious scientific discoveries the "always recurring and important factor ... is the simultaneity with which the complete solution is intuitively perceived and which can be checked later by discursive reasoning." She attributes the solution presented "as an archetypal pattern or image." As cited by von Franz, according to Jung, "Archetypes ... manifest themselves only through their ability to *organize* images and ideas and this is always an unconscious process which cannot be detected until afterwards."

Creativity and Affect

Some theories suggest that creativity may be particularly susceptible to affective influence.

Creativity and Positive Affect Relations

According to Isen, positive affect has three primary effects on cognitive activity:

1. Positive affect makes additional cognitive material available for processing, increasing the number of cognitive elements available for association.
2. Positive affect increases cognitive flexibility, increasing the probability that diverse cognitive elements will be fact become associated. Together,

these processes lead positive affect to have a positive influence on creativity.

3. Positive affect leads to defocused attention and a more complex cognitive context, increasing the breadth of those elements that are treated as relevant to the problem.

Fredrickson in her Broaden and Build Model suggests that positive emotions such as joy and love broaden a person's available repertoire of cognitions and actions, thus, enhancing creativity.

According to these researchers, positive emotions increase the number of cognitive elements available for association (attention scope) and the number of elements that are relevant to the problem (cognitive scope).

Creativity and Negative Affect Relations

On the other hand, some theorists have suggested that negative affect leads to greater creativity. A cornerstone of this perspective is empirical evidence of a relationship between affective illness and creativity. In a study of 1,005 prominent 20th century individuals from over 45 different professions, the University of Kentucky's Arnold Ludwig found a slight but significant correlation between depression and level of creative achievement. In addition, several systematic studies of highly creative individuals and their relatives have uncovered a higher incidence of affective disorders (primarily bipolar illness and depression) than that found in the general population.

Creativity and Affect at Work

Three patterns may exist between affect and creativity at work: Positive (or negative) mood, or change in mood, predictably precedes creativity; creativity predictably precedes mood; and whether affect and creativity occur simultaneously.

It was found that not only might affect precede creativity, but creative outcomes might provoke affect as well. At its simplest level, the experience of creativity is itself a work event and like other events in the organizational context, it could evoke emotion. Qualitative research and anecdotal accounts of creative achievement in the arts and sciences suggest that creative insight is often followed by feelings of elation. *For example,* Albert Einstein called his 1907 general theory of relativity "the happiest thought of my life." Empirical evidence on this matter is still very tentative.

In contrast to the possible incubation effects of affective state on subsequent creativity, the affective consequences f creativity are likely to be more direct and immediate. In general, affective events provoke immediate and relatively— fleeting emotional reactions. Thus, if creative performance at work-is an affective event for the individual doing the creative work, such an effect would likely be evident only in same-day data.

Another longitudinal research found several insights regarding the relations between creativity and emotion at work. First—a positive relationship between positive affect and creativity and no evidence of a negative relationship. The more positive a person's affect on a given day, the more creative thinking they evidenced that day and the next day even controlling for that next day's mood. There was even some evidence of an effect two days later

In addition, the researchers found no evidence that people were more creative when they experienced both positive and negative affect on the same day. The weight of evidence supports a purely linear form of the affect-creativity relationship, at least over the range of affect and creativity covered in our study: the more positive a person's affect, the higher their creativity in a work setting.

Finally, they found four patterns of affect and creativity affect can operate as an antecedent to creativity; as a direct consequence of creativity; as an indirect consequence of creativity; and affect can occur simultaneously with creative activity. Thus, it appears mat people's feelings and creative cognitions are interwoven in several distinct ways within the complex fabric of their daily work lives.

CREATIVITY AND INTELLIGENCE

There has been debate in the psychological literature about whether intelligence and creativity are part of the same process (the conjoint hypothesis) or represent distinct mental processes (the disjoint hypothesis). Evidence from attempts to look at correlations between intelligence and creativity from the 1950s onwards, by authors such as Barron, Guilford or Wallach and Kogan, regularly suggested that correlations between these concepts were low enough to justify treating them as distinct concepts.

Some researchers believe that creativity is the outcome of the same cognitive processes as intelligence and is only judged as creativity in terms of its consequences, *i.e.,* when the outcome of cognitive processes happens to produce something novel, a view which Perkins has termed the "nothing special" hypothesis.

A very popular model is what has come to be known as "the threshold hypothesis", proposed by Ellis Paul Torrance, which holds that a high degree of intelligence appears to be a necessary but not sufficient condition for high creativity. This means that, in a general sample, there will be a positive correlation between creativity and intelligence, but this correlation will not be found if only a sample of the most highly intelligent people are assessed. Research into the threshold hypothesis, however, has produced mixed results ranging from enthusiastic support to refutation and rejection.

An alternative perspective, Renzulli's three-rings hypothesis, sees giftedness as based on both intelligence and creativity. More on both the threshold hypothesis and Renzulli's work can be found in O'Hara and Sternberg.

Neurobiology of Creativity

The neurobiology of creativity has been addressed in the article "Creative Innovation: Possible Brain Mechanisms." The authors write that "creative innovation might require coactivation and communication between regions of the brain that ordinarily are not strongly connected". Highly creative people who excel at creative innovation tend to differ from others in three ways:

- They have a high level of specialized knowledge.
- They are capable of divergent thinking mediated by the frontal lobe.
- They are able to modulate neurotransmitters such as norepinephrine in their frontal lobe.

Thus, the frontal lobe appears to be the part of the cortex that is most important for creativity.

This article also explored the links between creativity and sleep, mood and addiction disorders and depression.

In 2005, Alice Flaherty presented a three-factor model of the creative drive. Drawing from evidence in brain imaging, drug studies and lesion analysis, she described the creative drive as resulting from an interaction of the frontal lobes, the temporal lobes and dopamine from the limbic system: The frontal lobes can be seen as responsible for idea generation and the temporal lobes for idea editing and evaluation. Abnormalities in the frontal lobe (such as depression or anxiety) generally decrease creativity, while abnormalities in the temporal lobe often increase creativity. High activity in the temporal lobe typically inhibits activity in the frontal lobe and vice versa. High dopamine levels increase general arousal and goal directed behaviors and reduce latent inhibition and all three effects increase the drive to generate ideas.

Working Memory and the Cerebellum

Vandervert described how the brain's frontal lobes and the cognitive functions of the cerebellum collaborate to produce creativity and innovation. Vandervert's explanation rests on considerable evidence that all processes of working memory (responsible for processing all thought) are adaptively modeled by the cerebellum. The cerebellum (consisting of 100 billion neurons, which is more than the entirety of the rest of the brain is also widely known to adaptively model all bodily movement. The cerebellum's adaptive models of working memory processing are then fed back to especially frontal lobe working memory control processes where creative and innovative thoughts arise. (Apparently, creative insight or the "aha" experience is then triggered in the temporal lobe.) According to Vandervert, the details of creative adaptation begin in "forward" cerebellar models which are anticipatory/ exploratory controls for movement and thought. These cerebellar processing and control architectures have been termed Hierarchical Modular Selection and Identification for Control (HMOSAIC). New, hierarchically arranged levels of the cerebellar control architecture (HMOSAIC) develop as mental mulling in working memory is extended over time. These new levels of the control architecture are fed forward to the frontal lobes. Since the cerebellum adaptively models all movement and all levels of thought and emotion, Vandervert's approach helps explain creativity and innovation in sports, art, music, the design of video games, technology, mathematics, the child prodigy and thought in general.

REM Sleep

Creativity involves the forming of associative elements into new combinations that are useful or meet some requirement. Sleep adds this process. REM rather than NREM appears to be responsible. This has been suggested to be due to changes in cholinergic and noradrenergic neuron-modulation that occurs during REM sleep. During this period of sleep high levels of acetylcholine in the hippocampus suppress feedback from hippocampus to the neocortex and lower levels of acetylcholine and norepinephrine in the neocortex encourage the spread of associational activity within neocortical areas without control from the hippocampus. This is in contrast to waking consciousness, where higher levels of norepinephrine and acetylcholine inhibit recurrent, connections in the neocortex. It is proposed that REM sleep would add creativity by allowing "neocortical structures to reorganize associative hierarchies, in which information from the hippocampus would be reinterpreted in relation to previous semantic representations or nodes."

Creativity and Mental Health

A study by psychologist J. Philippe Rushton found creativity to correlate with intelligence and psychotic sea. Another study found creativity to be greater in schizotypal' than in either normal or schizophrenic individuals. While divergent thinking was associated with bilateral activation of the prefrontal cortex, schizotypal individuals were found to have much greater activation of their *right* prefrontal cortex. This study hypothesizes that such individuals are better at accessing both hemispheres, allowing them to make novel associations at a faster rate. In agreement with this hypothesis, ambidexterity is also associated with schizotypal and schizophrenic individuals.

Particularly strong links have been identified between creativity and mood disorders, particularly manic-depressive disorder (a.k.a. bipolar disorder) and depressive disorder (a.k.a. unipolar disorder). In *Touched with Fire: Manic-Depressive Illness and the Artistic Temperament,* Kay Redfield Jamison summarizes studies of mood-disorder rates in writers, poets and artists. She also explores research that identifies mood disorders in such famous writers and artists as Ernest Hemingway (who shot himself after electroconvulsive treatment), Virginia Woolf (who drowned herself when she felt a depressive episode coming on), composer Robert Schumann (who died in a mental institution) and even the famed visual artist Michelangelo.

MEASURING CREATIVITY

Creativity Quotient

Several attempts have been made to develop a *creativity quotient* of an individual similar to the Intelligence Quotient (IQ), however these have been unsuccessful. Most measures of creativity are dependent on the personal judgement of the tester, so a standardized measure is difficult, if not impossible, to develop.

Psychometric Approach

J. P. Guilford's group, which pioneered the modem psychometric study of creativity, constructed several tests to measure creativity in 1967:

- Plot Titles, where participants are given the plot of a story and asked to write original titles.
- Remote Consequences, where participants are asked to generate a list of consequences of unexpected events (*e.g.*, loss of gravity)
- Quick Responses is a word-association test scored for uncommonness.
- Unusual Uses is finding unusual uses for common everyday objects such as bricks.
- Remote Associations, where participants are asked to find a word between two given words (*e.g.*, Hand—Call)
- Figure Concepts, where participants were given simple drawings of objects and individuals and asked to find qualities or features that are common by two or more drawings; these were scored for uncommonness.

Building on Guilford's work, Torrance developed the Torrance Tests of Creative Thinking in 1966. They involved simple tests of divergent thinking and other problem-solving skills, which were scored on:

1. Fluency. The total number of interpretable, meaningful and relevant ideas generated in response to the stimulus.

2. Originality. The statistical rarity of the responses among the test subjects.

3. Elaboration. The amount of detail in the responses. The Creativity Achievement Questionnaire, a self-report test that measures creative achievement across 10 domains: Was described in 2005 and shown to be reliable and valid: When compared to other measures of creativity and to independent evaluation of creative output.

Social-Personality Approach

Some researchers have taken a social-personality approach to the measurement of creativity. In these studies, personality traits such as independence of judgment, selfconfidence, attraction to complexity, aesthetic orientation and a risk-taking are used as measures of the creativity of individuals. Other researchers have related creativity to the trait, *openness to experience.*

Other Approaches to Measurement

Genrich Altshuller in the 1950s introduced approaching creativity as an *exact science* with TRIZ and a Level-of-Invention measure.

The creativity of thousands of Japanese, expressed in terms of their problem-solving and problem-recognizing capabilities, has been measured in Japanese firms.

Howard Gruber insisted on a case-study approach that expresses the existential and unique quality of the creator. Creativity to Gruber was the product of purposeful work and this work could be described only as a confluence of forces in the specifics of the case.

CREATIVITY AND VARIOUS CONTEXTS

Creativity has been studied from a variety of perspectives and is important in numerous contexts. Most of these approaches are undisciplinary and it is therefore difficult to form a coherent overall view. The following sections examine some of the areas in which creativity is seen as being important.

Creativity in Diverse Cultures

Creativity is a scientific concept which is mostly rooted within a Western creationist perspective. Francois Jullien in 'Process and Creation, 1989' is inviting us to look at that concept from a Chinese cultural point of view. Fangqi Xu has reported creativity courses in a range of countries. Todd Lubart has studied extensively the cultural aspects of creativity and innovation.

Creativity in Art and Literature

Most people associate creativity with the fields of art and literature. In these fields, *originality* is considered to be a sufficient condition for creativity, unlike other fields where both *originality* and *appropriateness* are necessary.

Within the different modes of artistic expression, one can postulate a continuum extending from "interpretation" to "innovation". Established artistic movements and genres pull practitioners to the "interpretation" end of the scale, whereas original thinkers strive towards the "innovation" pole. Note that we conventionally expect some "creative" people (dancers, actors, orchestral members, etc.) to perform (interpret) while allowing others (writers, painters, composers, etc.) more freedom to express the new and the different.

Contrast alternative theories, *for example*:

- Artistic inspiration, which provides the transmission of visions from divine sources such as the Muses; a taste of the Divine. Compare with invention.
- Artistic conversation, as in Surrealism, which stresses the depth of communication when the creative product is the language.
- Artistic evolution, which stresses obeying established' ("classical") rules and imitating or appropriating

to produce subtly different but unshockingly understandable work. Compare with crafts.

In the art practice and theory of Davor Dzalto, human, creativity is taken as a basic feature of both the personal existence of human being and art production. For this thinker, creativity is a basic cultural and anthropological category, since it enables human manifestation in the world as a "real presence" in contrast to the progressive "virtualisation" of the world.

Creative Industries and Services

Today, creativity forms the core activity of a growing section of the global economy—the so-called "creative industries"—capitalistically generating (generally non-tangible) wealth through the creation and exploitation of intellectual property or through the provision of creative services. The Creative Industries Mapping Document 2001 provides an overview of the creative industries in the UK The creative professional workforce is becoming a more integral part of industrialized nations' economies.

Creative professions include writing, art, design, theater, television, radio, motion pictures, related crafts, as well as marketing, strategy, some aspects of scientific research and development, product development, some types of teaching and curriculum design and more. Since many creative professionals (actors and writers, *for example*) are also employed in secondary professions, estimates of creative professionals are often inaccurate. By some estimates, approximately 10 million US workers are creative professionals; depending upon the depth and breadth of the definition, this estimate may be double.

Creativity in Other Professions

Creativity is also seen as being increasingly important in a variety of other professions. Architecture and industrial design are the fields most often associated with creativity and more generally the fields of design and design research.

These fields explicitly value creativity and journals such as *Design Studies* have published many studies on creativity and creative problem solving.

Fields such as science and engineering have, by contrast, experienced a less explicit (but arguably no less important) relation to creativity. Simonton shows how some of the major scientific advances of the 20th century can be attributed to the creativity of individuals. This ability will also be seen as increasingly important for engineers in years to come.

Accounting pas also been associated with creativity with the popular euphemism *creative accounting.* Although this term often implies unethical practices, Amabile has suggested that even this profession, can benefit from the (ethical) application of creative thinking.

Creativity in Organizations

Amabile argued that to enhance creativity in business, three componerits were needed:

- Expertise (technical, procedural & intellectual.
- Motivation (especially intrinsic motivation).
- Creative thinking skills (how flexibly and imaginatively people approach problems),

Nonaka, who examined several successful Japanese companies, similarly saw creativity and knowledge creation as being important to the success of organizations. In' particular, he emphasized the role that tacit knowledge has to play in the creative process. '

Economic Views of Creativity

In the early 20th century, Joseph Schumpeter introduced the economic theory of *creative destruction,* to describe the way in which old ways of doing things are endogenously destroyed and replaced by the new. Creativity is also seen by economists such as Paul Romer as an important element in the recombination of elements to produce new technologies and products and, consequently, economic growth. Creativity leads to capital and creative products are protected by intellectual property laws.

Creativity is also an important aspect to understanding Entrepreneurship.

The *creative class* is seen by some to be an important driver of modern economies. In his 2002 book, *The Rise of the Creative Class,* economist Richard Florida popularized the notion that regions with "3 T's of economic development:

Fostering Creativity

Daniel Pink, in his 2005 book A Whole New Mind, repeating arguments posed throughout the 20th century, argues that we are entering a new age where creativity is becoming increasingly important. In this conceptual age, we will need to foster and encourage right-directed thinking (representing creativity and emotion) over left-directed thinking (representing logical, analytical thought).

Nickerson provides a summary of the various creativity techniques that have been proposed. These include approaches that have been developed by both academia and industry:

1. Establishing purpose and intention.
2. Encouraging acquisitions of domain-specific knowledge.
3. Building basic skills.
4. Encouraging confidence and a willingness to take risks.
5. Focusing on mastery and self-competition.
6. Stimulating and rewarding curiosity and exploration.
7. Building motivation, especially internal motivation.
8. Developing self-management (metacognitive skills).
9. Teaching techniques and strategies for facilitating creative performance.
10. Promoting supportable beliefs about creativity.

11. Providing opportunities for choice and discovery.
12. Providing balance.

Some see the conventional system of schooling as "stifling" of creativity and attempt (particularly in the preschool/kindergarten and early school years) to provide a creativity-friendly, rich, imagination-fostering environment for young children. Compare Waldorf School.

A growing number of psychologists are supporting- the idea that there are methods of increasing the creativity of an individual. Several different researchers have proposed approaches to prop up this idea, ranging from psychological-cognitive, such as:

- Osborn-Parnes Creative Problem Solving Process.
- Synectics.
- Inventium and science-based creative thinking.
- Purdue Creative Thinking Program.

and

- Lateral thinking (courtesy of Edward de Bono), to the highly-structured, such as:
- TRIZ (the Theory of Inventive Problem-Solving);
- ARIZ (the Algorithm of Inventive Problem-Solving), both developed by the Russian scientist Genrich Altshuller.
- Computer-Aided Morphological analysis.

Creative Process with New Technologies

A simple but accurate review on this new Human-Computer Interactions (HCI) angle for promoting creativity has been written by Todd Lubart, an invitation full of creative ideas to develop further this new field.

Groupware and other Computer Supported Collaborative Work (CSCW) platforms are now the stage of Network Creativity on the web or on other private networks. These tools have made more obvious the existence of a more connective, co-operative and collective nature of creativity rather than the prevaling individual one. Creativity Research on Global Virtual Teams is showing that the creative process is affected by the national identities, cognitive and conative profiles, anonymous interactions at times and many other factors affecting the teams members, depending on the early or later stages of the cooperative creative process. They are also showing how NGO's cross cultural virtual team's innovation in Africa would also benefit from the pooling of best global practices online. Such tools enhancing 'cooperative creativity may have a great impact on society and as such should be tested while they are built following the Mottos, "Build the Camera while shooting the film". Some European FP7 scientific programs like Paradiso are answering a need for advanced experimentally-driven research including large scale experimentation test-beds to discover the technical, societal and economic implications of such groupware and collaborative tools to the Internet.

On the other hand Creativity research may one day be pooled with a computable Metalanguage like IEML from the University of Ottawa Collective Intelligence Chair, Pierre Levy. It might be a good tool to provide an interdisciplinary definition and a rather unified theory of creativity. The creative processes being highly fuzzy, the programming of cooperative tools for creativity and innovation should be adaptive and flexible. Empirical Modelling seems to be a good choice for Humanities Computing.

If all the activity of the universe could be traced with appropriate captors, it is likely that one could see the creative nature of the universe to which humans are active contributors. After the web of documents, the Web of Things might shed some light on such a universal creative phenomenon which should not be restricted to humans. In order to trace and enhance co-operative and collective creativity, Metis Reflexive Global Virtual Team has worked for the last few years on the development of a Trace Composer at the intersection of personal experience and social knowledge.

Metis Reflexive Team has also identified a paradigm for the study of creativity to bridge European theory of "useless" and non instrumentalized creativity, North American more pragmatic creativity and Chinese culture stressing ·more creativity as a holistic process of continuity rather than radical change and originality. This paradigm is mostly based on the work of the German philosopher Hans Joas, one that emphasizes the creative character of human action. This model allows also for a more comprehensive theory of action. Toas elaborates some implications of his model for theories of social movements and social change. The connexion between concepts like creation, innovation, production and expression is facilitated by the creativity of action as a metaphore but also as scientific concept. The Creativity and Cognition conference series, sponsored by the ACM and running since 1993 has been an important venue for publishing research on the intersection between technology and creativity. The conference now runs biannually: next taking place in 2009.

Social Attitudes to Creativity

Although the benefits of creativity to society as a whole have been noted, social attitudes about this topic remain divided. The wealth of literature regarding the development of creativity and the profusion of creativity techniques indicate wide acceptance, at least among academics, that creativity is desirable.

There is, however, a dark side to creativity, in that it represents a *"quest for a radical autonomy apart from the constraints of social responsibility"*. In other words, by encouraging creativity we are encouraging a departure from society's existing norms and values. Expectation of conformity runs contrary to the spirit of creativity. Sir Ken Robinson argues that the current education system is "educating people out of their creativity"

Nevertheless, employers are increasingly valuing creative skills. A report by the Business Council of Australia, *for example,* has called for a higher level of creativity in graduates. The ability to "think outside the box" is highly sought after. However, the above-mentioned paradox may well imply that firms pay lip service to thinking outside the box while maintaining traditional hierarchical organization structures in which individual creativity is not rewarded.

CREATIVITY TECHNIQUES

Creativity techniques are methods that encourage original thoughts and divergent thinking. Some techniques require groups of two or more people while other techniques can be accomplished alone. These methods include word games, written exercises and different types of improvisation. Creativity techniques can be used to develop new materials for artistic purposes or to solve problems.

Most creativity techniques use associations between the goal (or the problem), the current state (which may be an imperfect solution to the problem) and some stimulus (possibly selected randomly).

There is an analogy between many creativity techniques and methods of evolutionary computation.

Randomness

Randomness, or aleatory, is the introduction of chance elements. Aleatory is commonly found in music, art and literature, particularly in poetry. In film andy Voda made a movie in 1979 called "Chance Chants" which he produced by a flip of a coin, or roll of a dice. In music, John, Cage, an avant-garde musician, composed music by superimposing star maps on blank sheet music, by rolling dice and by preparing open ended scores that depended on the spontaneous decisions of the performers. (1) Other ways of practicing randomness include coin tossing, picking something out of a hat, or selecting random words from a dictionary.

In short, aleatory is a way to introduce new thoughts or ideas into a creative process.

Improvisation

Improvisation is a creative process which can be spoken, written, or composed without prior preparation. Improvisation, also called extemporization, can lead to the discovery of new ways to act, new patterns of thought and' practices, or new structures.

Improvisation is used in the creation of music, theatre and other various forms. Many artists also use improvisational techniques to help their creative flow. Here, are two significant methods:

> Improvisational Theater is a form of theater in which actors use improvisational acting techniques to perform spontaneously. Many "improve" techniques are taught in standard drama classes. The basic skills of listening, clarity, confidence and performing instinctively' and spontaneously are considered important skills for actors to develop.

Free Improvisation is real time composition. Musicians of all kinds "improv" music; this music is not limited to particular genre. Two contemporary musicians that use free improvisation are Anthony Braxton and Cecil Taylor. Through free improvisation, musicians can develop increased spontaneity and fluency.

Each type of improvisation improves the thinking and acting skill of the actor, this is made by using no practise, a similar set of techniques is called *Alienation* since one 'of its many techniques uses actors that haven't rehearsed or even read the play, improvisation is an acting skill where actors make up a storyline, start and ending on the spot and actors have to try their best to keep in character.

Problem Solving

In problem-solving contexts, the random word creativity technique is perhaps the simplest method. A person confronted with a problem is presented with a randomly generated word, in the hopes of a solution arising from any associations between the word and the problem.

A random image, sound, or article can be used instead of a random word as a kind of creativity goad or provocation.

Therapeutic Applications

Projective therapy is a form of psychotherapy which uses synthesis to produce original interpretation and problem solving strategies. Fiction is read in the form of a novel or observed in the form of a film. In this therapeutic technique, patients analyze the fictional work and immerse themselves in this fictional world. Rather than focusing on their problems they analyze the author- or actors- motivation, plot-development and cinematography during the group process. By reading/viewing many consecutive works during this process, the patients are able to achieve a creative link between their own issues and those of the fictional characters. This synthesis is usually unique to each individual since it bypasses the conscious and directly accesses the unconscious by linking with semiotic objects and archetypes which populated this otherwise inaccessible part of the mental process. This technique was first used by Dr. A. James Giannini at Yale University using the World of Tiers science-fiction series written by Philip Jose Farmer. A later novel in this series, Red arc's Rage, was produced by both Farmer and Giannini. This particular novel reviewed the technique of projective therapy; the use of the creativity unleashed in this process: And the evolution of the' Tiers series itself.

❋❋❋

7

Motivation and Group Dynamics

It is an accepted fact that behaviour of an individual is very much guided by instincts and accompanying emotions. In the present set-up, there is also a school of psychologists who do not accept the instincts as the cause of the behaviour of the individual. Psychologists of this school put forward a theory that the behaviour of an individual is guided by motive and drives. Here we shall try to discuss about all these things.

According to H. J. Klanmesier, "Motivation is an integral component of learning and a critical problem in educational settings. Educational psychologists and educators use such terms as needs, interests and goals in treating the motivational aspects of teaching learning processes; in other specialities in psychology primary drives, secondary drives and incentives appear frequently."

According to M. Jolmson, "Motivation is the influence of general pattern of activity indicating and directing the behaviour of the organism." "The influence of the general pattern of activity, initiating and directing the behaviour of the organism, goes by the name of motivation."

According to N.L. Munn, "The inner control of behaviour as represented by physiological conditions, interests, attitudes and aspirations."

Meaning of Drives

As mentioned above, R. S. Woodworth suggested the term "drive" for urges. In the beginning 'drive' meant the energy that sets things going. But lateron, its meaning was expanded and "drive" stood for "energy" or action arising from physiological needs like hunger, the rot, sex, fatigue, elimination of bodily wastes, maintenance of constant temperature and barometric pressures." L. F. Shaffer has defined drive as a "stineubes, usually internal, that arouses persistent mass activity" and a mechanism as "an acquired response tending to the reduction of a drive."

On the basis of these definitions Shaffer point out that many drives may be innate and all mechanisms are the results of learning processes. But as we have seen, Woodworth has a slightly different view, for according to him every mechanism has the power to become a drive. Woodworth stated that drives can be acquired also. That is why, some psychologists suggest that there are primary and secondary drives. Primary drives are nature and secondary drives are acquired.

Element of the Drive

There are following three elements that drive or stimulate persons to take to some work:

1. The excitement, which inspires a person to do some work.
2. This power is stimulated from within.
3. The hope, that if the work has been completed there shall be some peace and satisfaction.

Causes of Motivation

Following are the various causes that bring about motivation:

1. He feeling of safe-guarding against pain and unhappiness.
2. The desire of the person to survive and live.
3. Internal love or affection.

4. Desire for pleasure.
5. He feeling of self-preservation or security.
6. Sex instinct.
7. Emotions, instinct, thinking, feeling, etc.
8. Sub-conscious mind.
9. Power of determination of desire to do a thing. It may also be termed as *Will*.
10. Various circumstancial or environmental factors that direct the movement of the body.

Importance of Motivations in Education

Motivations play a vital role in the field of education. M. K. Thomson is of the opinion that motivation is that education, by which interest is created in a person, who does not have it. Due to interest the students concentrate their attention on studies. If an interest is created for a particular subject in the child, then he shall learn it easily and quickly. By motivation, the development of the character and the personality of the child is also possible. In short it can be said that motivations have important place in education in this regard, it is necessary to see to the following:

(a) The teacher should try to develop the spirit of co-operation in the students.

(b) The teacher should try to create group feeling or the feeling of working and living together, amongst the students.

(c) The strongest motivation of the child should be awakened to the maximum.

(d) So far as possible, the teacher should try to awaken the motivation.

(e) The teacher should try to present good examples before the students. These example shall inspire the children.

(*f*) The teacher should be enthusiatic and zealous. Every individual has an aim or goal. It is necessary to make efforts to reach that goal. The students should be encouraged to set a goal and make purposeful attempt to reach it.

CHARACTERISTICS OF MOTIVATED BEHAVIOUR

The analysis of place of motive in experience and behaviour as given above, clarifies the general characteristics of motivated behaviour. Roughly speaking, these characteristics are as follows :

1. Energy Mobilization. Even the most peaceful animal becomes frightening in times of emergency. This is due to the mobilisation of energy. This process of the mobilisation of energy is chemical as well as physical. It cannot be perceived directly, but when there is some difficulty in the instilling in him love towards good habits, hatred for bad ones, anger for injustice, curiosity towards nature, etc. Once there emotions are conjoined to the right object, his behaviour takes a good turn. Sublimation does not destroy the emotion, it only changes the mode of its expression. It achieves the necessary control while avoiding all the damage that is the inevitable corollary of repression. It also avoids the formation of emotional complexes. As psychoanalytic studies have revealed, most of the great poets, authors, scientists, leaders and saints could contribute to human well being only through the sublimation of their emotions. The compositions and creations of both *Tulsidas* and *Kalidas* clearly show the results of sublimation of their love for the opposite sex, as a result of which their sexual instinct was sublimated into love of God.

Attainment of the object of motive, it is seen in the form of extra energy. The animal tries to attain the object with as much power as is the urge of the motive behind it. A hungry dog takes the bread even while beaten. The labour which a student puts in to attain a first division in the examination is never seen in his ordinary life. Thus, it is often observed that some of the persons brought up in the poor families are very persevering and laborious. Some men perform such inhuman action due to the sex urge which they would never do in ordinary conditions.

2. Persistence. An animal cannot leave its work without achieving the goal of its motive. An animal wandering in search of food keeps on roaming so long as it does not get it. The fighting birds keep on fighting till either of them is wounded or runs away. Sometimes some motivated actions go on for years. A young man in search of a job goes on labouring for several years till he gets a suitable one. A lover determined to conquer his beloved, does not leave her even in the face of great difficulties. Even in the motivated behaviour of insects and animals much patience and persistence is observed.

3. Restlessness for the Attainment of the Goal. the motivated behaviour is goal-directed. The restlessness observed in such a behaviour persists so long as the goal is not achieved. *For example,* the goal of hunger is food. Till the food is taken the restlessness disappears. The goal can be conscious or unconscious. Sometimes the man knows the goal of his motive, but at other times it is only after long efforts and trial and error that he knows the real object which will bring his restlessness to an end. The physical and the mental motives have physical and mental goals respectively.

4. Variability. Motive has a goal, hence, it continuously changes with the veiw of attainment of goals. A hungry cat closed in a cage tries new devices to open it. Sometimes it tires this way, sometimes that and this change of behaviour goes on till it succeeds in coming out of the cage. A man with the goal of earning money tries hundreds of ways for it. Sometimes he purchases a lottery ticket while at other times he is busy in some business. In fact, the motivated person goes on changing his ways till he arives at the way successful for the achievement of the aim.

5. Extinction of Restlessness After the Attainment of the Goal. The restlessness of the motivated behaviour exhibits itself in following ways of the fulfillment of some internal want or desire and extinguishes as soon as that want or desire is fulfilled. A thirsty animal roams in search of water. As it gets water it drinks it and the restlessness is gone. There are only two ways to get rid of the restlessness of the motive. Either the tension should be removed or it should be changed. The tension can be removed by the attainment of the object. Sometimes conflicts are observed due to tensions between more than one motives at the same time.

TYPES OF MOTIVATION

M. K. Thomson has classified motivation under two heads:

A. Natural Motivation

Natural Motivation is of the following five types:

1. Bodily Goals or Motivation. These motivations are nothing but bodily or physical needs, like hunger, thirst, sex, *etc.* Certain motivations are caused by physical activities such as respiration, throbbing of the heart, digestion, circulation of blood and working of the glands etc. These motivations are inborn and inspire a person to fulfil the needs of the body. By causing change in the glands of a person, it is possible to bring him to the normal situation. Such a situation will be helpful for the information of motivation.

Various experiments that have been done, put forward the conclusion, that in future, by physical and operational experiments and chemical activities, it shall be possible to bring about changes in the glands of the person. Such a change shall give rise to requisite of natural and required motivations.

2. Habit. Habit is a form of nature of a person. A good habit of a child shall be that which can bring about development in his character and help him in achievement of education.

3. Instinct Motivation. Instincts, as already said, are inborn powers and they inspire the physical organs of the individual to take to certain jobs. Unless a person has satisfied his instincts, he remains restless. Teachers, in order to provide for their satisfaction, should present a normal atmosphere for the students.

4. Feelings and Emotions. Pain and pleasure, joy and distress are the natural feelings and experiences of human nature. This aspect of the nature and the personality is very helpful in that system of education. Where an atmosphere is aimed to be created, which can bring about the development of the character of the students and help them to educate themselves. This aspect of the personality should be very cautiously handled, otherwise, it can have adverse effect.

5. Emotional Motivations. Those motivations, that inspire us to bring about the fulfilment of certain emotional needs, are called "Emotional Motivations". Emotions are awakened by the outside stimuli. These stimuli are part of the environment. The teachers should try to create such a healthy environment that the sublimation or motivation of the emotions may be possible.

Other than the above five Main Motivations, there are certain other motivations as well. These motivations are also part of natural motivations. It shall be wise to have an idea about them as well.

1. Social Motivations. Due to inspirations from social environment, a person takes to certain activities. These activities are guided by the emotions and the instincts. Normally certain motivations are awakened for the fulfilment of social needs. They express themselves in the following forms:

(a) Gregariousness is a natural instinct of man. It is therefore natural for him to be guided by social motivations.

(b) The man has a great social consciousness of getting a place of respect in the society or to get his Self-

regarding sentiment fulfilled. Consequently he tries to do certain good things in the society.

(*c*) Due to social motivations, a person takes up those jobs that are good from the point of view of the society or which are appreciated by the society.

2. Desires. It is the duty of the teachers to create in the students, the desire to acquire knowledge and achieve success. Is the teacher succeed it is, he shall be achieving a great task.

3. Individual Motivations or Motivations of Self. Those motivations, that inspire a person to satisfy his personal needs are called individual motivations, beauty, knowledge, faithfulness, competition, ideal, hero worship etc., are personal needs. These needs inspire a person to fulfill his personal desires.

4. Recreation and Play. Programmes of games and play are joyful for the students. In plays, elements of direct motivation are present. The elements can therefore, by very nicely utilised for imparting education to the students.

5. Direction and Imitation. Children have the tendency to imitate. If proper direction and examples are presented before them, they can improve the power of motivation. They can also lead them towards success.

6. Appreciation and Self-expression. It is natural for a man to be inclined towards such goods goals, that give him importance and greatness. Social appreciation is another form of this feeling. It is, therefore, necessary to provide students the opportunity to observe these things.

7. Other Natural Motivations. Other than these things, man is also anxious to have knowledge, love, faith, office, ideal, rest, leadership, achievement of wealth etc. It is quite possible to utilise all these motivations of the child in the field of education. It is clear, that in human nature there is enough material for Natural Motivation.

B. Artificial Motivations

Artificial motivations are of various types. These motivations are not present in the human personality. They are part of the environment or the society. It is useful for the teacher to know and use them. They are being given below:

1. Reward and Punishment. In the works of the school this serves as a very imiportant and useful motivation. It has the following accept as well:

(a) Punishment is a negative motivation. Sometimes it is not very useful for children.

(b) Reward is a positive motivation. Reward may be physical, social, or normal. It is very dear to the students. Teacher should try to use this motivation in such a way that the students may grow into useful future citizens.

2. Co-operation. It is also a strong motivation. The teacher can use it as a medium of education. In the Project Method of Education attempt is made to develop a spirit of co-operation amongst the students. This motivation can be well utilised.

FUNCTIONS OF MOTIVATIONS

In modern educational psychology every possible effort is made to base the education of the child on its motivations. Modern education aims at the excitation of one or the other motivation in a child in order to make any and every kind of education natural and normal. Various motivations in a child are useful in education in the following ways:

1. Escape. In a child one finds the motivation of escape conjoined to the emotion of fear. The intelligent teacher arouses fear and a disgust towards bad things in the child. He also tries to prevent fear and escape from action in many unnecessary cases. Meaningless fears are thus, eliminated.

2. Pugnacity. The instinct of pugnacity of combat is a universal motive. The most important fact that accelerates or excites it is restraint. A child whose means of locomotion are stopped by the simple expedient of being placed under restraint will go red in the face and cry at the top of its voice. Conflict starts the moment one removes a chick from the hen, or makes fun of an irritable individual or prevents an animal from fulfilling and satisfying its motivations.

The object of animal in turning pugnacious is to remove the obstacle preventing its progress or to defeat the obstacle. As long as the individual fails to defeat the obstacle or to remove it from his path, his internal, physiological and psysiological disturbance continues. The pattern of pugnacity or combat is not the same in all animals or individuals nor are the implements common to all. Animals having long nails or teeth use them in warfare. In most animals it is the male who is pugnacious. He combats the animals of different species and other males of his own species for supremacy over the females or for food. It is possible that the instinct of the pugnacity is somewhat depressed by the female hormones. Human beings and some other animals are known to threaten before actually attacking. Proverbially it is the monkey who is the best exponent of this behaviour pattern. Some animals threaten by showing their teeth, others by making offensive noises while others show their ungentlemanly instants by hanging on their chests. Forms of pugnacity are modified by age, intelligence, social and cultural influences and social behaviours. The pugnacious motivation of cultured individual finds expression is more indirect manner. His animosity continues over the years and his vengeance is achieved in many devious and indirect ways. Litigation is only a form of conflict and it is very common in cultured societies. In fact, it is difficult to believe that pugnacity is only an instinct or an internal motive whose single object is conflict or fight. The motivation to competition may be innate but the pugnacious motivation is activated only when the fulfilment of a motive is obstructed. Anger results in conflict but sometimes anger finds expression in more devious and uncommon ways. An employee who has been severely reproved by his employer is in no position to answer him back in strong language as he goes in fear of losing his job, but as an alternative he lets off his pent up feelings of anger on his wife and children. Sometimes our pugnacious instinct is satisfied when we see others fighting. The tension of conflict is also released and relieved when one plays games or sees others playing or fighting going on the film screen. Within the family the parents prevent many of the child's instinctive actions so that he develops feelings of animosity them. Children worry the parents so that the latter become angered. Children compete among themselves for the love of their parents.

3. Parental Motivation. To this motivation is conjoined the emotion of pity. This motivation finds its expression in the older child's love for his younger counterpart and for other living beings.

4. Repulsion. The intelligent teacher does his best to develop the motive of repulsion in the child for unsavoury and undesirable expendient.

5. Curiosity. Curiosity is the most important motivation from the educational viewpoint. It is a fundamental motivation without any definite biological basis. When any living being is confronted with any new or unrecognised object, it is natural for it to desire a complete knowledge or understanding of it. Curiosity in human beings is accompanied by the sense of surprise. Almost all individuals are desirous of knowing new things and new places. When a child is handed something it has not seen before, it will immediately turn it over and explore it to the best of its ability. The child is always ready with an unending list of questions. He never tires of asking questions even if the parents tire of answering them till he is forced to rebuff the child. If the child's curiosity is not suppressed it continues its exploration and questioning till it finally comes to know its surroundings to its own satisfaction. It is such children who turn into the best investigators and scientists when they maintain their curiosity intact in their adult circumstances.

6. Sex. Sex motivation of a person may be turned into more acceptable outlets such as dancing, music, art, etc., by the teacher. And, particularly in the adolescent stage, this motivation can be restricted and moulded to provide the desirable development in the adolescent.

7. Self-Assertion. The motivation of self-assertion occupies an importance place in the child's character. To it is conjoined the sentiment of self-respect which helps the child steer clear of all undersirable activity and to concentrate only on the desirable one. In order to develop self-respect, in the child, the teachers must put before him examples taken from the child's own environment. If instead of punishing a child for his mistake, the teacher tells him that such foolishness is not expected of him or from one of his upbringing, then the child will abstain from repeating it. He will also be remorseful. Once the motivation of self-assertion is fully developed, the individual makes his own efforts to progress in life, since he does not want to live the life of an insect but wants to hold his head high in society.

8. Gregariousness. A child is gregarious by motivation and he naturally wants to remain with the group of friends. A child does not desire solitude. The teacher can make the best of this instinct and help in the social development of the child. The teacher should keep an eye on the various group tensions that develop within the school and also try to evolve healthy competition, co-operation and team spirit in the children. Blind faith in small groups may be turned to more productive channels when it is turned to love of country and philanthropy. In this way, the child becomes socially mature.

9. Self-Submission. Of course, the motivation of self-submission may prove to be a hindrance in the child's development but the sensible teacher can turn it towards himself and arouse feelings of respect for the teacher and a healthy regard for discipline in the child. Students are prone to accept teacher's mandate without questioning so that success in education becomes possible.

10. Food Seeking. The food seeking motivation may be put to much fruit if indirect use by the teacher as he can instigate the student to greater effort in his study by inspiring him with vision of success in his future livelihood. Young men are seen to work very hard as they are afraid for their bread at a later stage. The motivation of self-assertion also contributes to this process.

11. Acquisitiveness. A child motivity collects many odds and takes great pleasure in doing it. It gives him a feeling of possession and pride. In their tender age children are often seen collecting stamps, match boxes, cirgarette boxes and are often seen fighting to the better end for them. Within the bounds of the school, this innate tendency of acquisition can be put to constructive use. They can be guided to collect stamps, photographs, old coins and books etc. But in this process it is essential to see to it that the child does not resort to undersirable methods of augmenting his collection. In the school, the child's acquisitive tendency can be adequately exploited to create a small library or a collecting of odd bits, in which the child can retain possession over them and yet allow them to be used by all his colleagues.

12. Laughter. The motivation to laughter is accompanied by the emotion of enjoyment and relief. If the children are made to laugh in class they are more inclined to accept the teacher's suggestions. Hence, the sensible teacher can get the children to perform different tasks by arousing laughter and enjoyment.

TECHNIQUES OF MOTIVATING STUDENTS

1. Set Induction. It means the introduction of the lesson. It links previous knowledge with the present knowledge. It is known as the skill of introduction or set-induction skill.

2. Stimulus Variation. It means changing of gestures and positions by the teacher. If a teacher does not change his gestures and positions during the teachning process, it becomes bore and lacks in interest. Hence, it is necessary to provide the training to the teachers in the skill of changing the gestures.

3. Lecture. It is concerned with the effective presentation of the content. The teacher leaves his impressions by using many techniques and tactics through this skill. It is also known as 'Communication skill'.

4. Probing Questions. it is concerned with the questions to be asked about the content in more depth. It stimulates the cognitive development of the pupils.

5. Illustration. There are two teaching methods-continuous lecturing method and demonstration method. The pupil-teachers should explain the concepts through examples and by displaying pictures and charts. It is called the illustrating skills.

6. Closure. It means to finish some task, *i.e.,* in class the pupil-teacher exhibits various behaviours. If we divide these behaviours in smaller units, these are termed as 'skills'. When a pupil-teacher delivers lecture and sums up properly and in an attractive way, the skill is termed as 'Closure Skill'. The lesson remains ineffecitve in the absence of proper closure.

7. Increasing Pupil Participation. It is concerned with increasing pupil participation which means—'Pupils' direct behaviour which is observable. This includes both responses and reactions of the pupils along with their own new activities.

8. Use of Black Board. It is very essential in the class. Its use also needs special training. The necessary components of blackboard work are clarity of handwriting, legibility and rationale of blackboard work etc.

9. Skills for Class Management. Both—social as well as educational activties performed in order to create proper environment for learning in the classroom. The performance of these activities needs special skill. As these activities manage the class it is called 'Skill for Class Management'.

10. Skill of Explaining. It means—use of explaining or connecting links to link the statements or systematic information. When a teacher shows his behaviour while explaining the pupils about 'What', 'Why' and 'How' regarding some facts, principles and concepts, that behaviour constitutes the skill of explaining.

MOTIVATION (CAUSED BEHAVIOUR)

Motivation is essential to the operation of organizations and classroom activities. No matter how much equipment and audio-visual aids a teacher has, these things cannot be made use of unless students are motivated. From psychology we learn that normal behaviour is caused by certain reasons. These reasons may relate to a person's needs and or the consequences that result from acts. Motives are expressions of a person's needs. Hence, they are personal and internal, Incentives on the other hand, are external to the person.

There are various ways to classify needs. A simple but significant one is *(1)* Basic physiological or primary needs and *(2)* Social and psychological needs, called secondary needs.

Conscious Action Arises from Needs

Diagrammed very simply, the relation of needs to action is shown in the figure below. Needs create tensions that are modified.

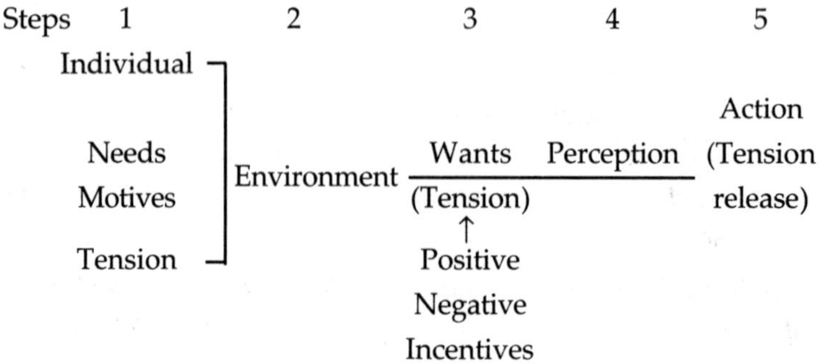

1. Physiological Needs. Some of the physiological needs are food, water, sex, sleep, air to breathe and satisfactory temperature. These needs arise out of the basic physiology of life and are important to survival and preservation of the species.

2. Secondary Needs. Secondary needs represent needs of the mind and spirit, rather than of the physical body. They develop a variety of motives in each person. Many of these needs are developed as one matures. Examples are rivalry, self-esteem, sense of duty, self-assertion, giving, belonging and receiving affection. The secondary needs are the ones that complicate the motivational efforts of teachers.

Characteristics of Secondary Needs

1. They are strongly conditioned by experience.
2. They are often hidden from conscious recognition.
3. They work in groups rather than alone.
4. They vary in type and intensity among people.
5. They change within an individual.
6. They are nebulous feelings instead of tangible physical needs.
7. They influence behaviour. It is said, "We are logical only to the extent our feelings let us be."

Although human needs have been classified as primary and secondary, they are inseparable in a particular human being. The state of the physical body affects the mind and the state of mind can affect the physical body. In fact, medical research shows that both chemistry (a physical condition) affects the emotions and feelings. According to the renowned psychiatrist, Dr. William C. Menninger, "The psychiatrist believes that the mind and body are inseperable—the holistic concept. The former tendency to regard the mind and body as separate has been discarded. Now we study the physical, the chemical, the psychological and the social factors of every individual who comes to us for help."

The Expectancy Model of Motivation

Motivation in the expectancy model developed by Victor H. Vroom, Porter and Lawler and others, has been defined as the strength of drive toward action. The model shows that a person's motivation toward an action at a particular time is determined by the outcomes (valence) of the action, multiplied by the strength of a person's expectation that the action will lead to the outcomes. The expectancy model shows that motivation leads to increased effort. The increased effort will lead to outcomes that represent desired goals.

Valence × Expectancy = Motivation.

Valence refers to the strength of a person's preference for one's outcome in relation to others.

Expectancy. It is the strength of belief that a particular act will be followed by particular outcomes.

A Diagram of the Expectancy Model

Valence × Expectancy

↓

Motivation

↓

Action

↓

Outcomes (Goal Attainment)

Satisfaction

Various Theories of Motivation

Following theories of motivation explain motivation with all its distinct varieties. We may find that these theories supplement one another and point towards the same truth.

1. Pawn Theory. This is based on the transcendental approach to the problems of life. According to this theory, we are a 'pawn', a puppet, an instrument in the hands of God. Both classical and Freudian Psychologists land Behaviouristic theories of learning derive their strength from this. This pawn theory passes on the responsibility to some 'mysterious power' which is something intengible.

2. Instinct Theory. McDougall is said to be the originator of this theory. According to him, "The "human mind has certain innate or inherited tendencies which are the essential springs or motive powers of all thought and action, whether individual or collective and are the bases from which the character and will of individuals and of nations are gradually developed under the guidance of the intellectual faculties." McDougall put forward a list of 41 instincts and attached 14 emotions with them. This theory was very popular in Britain with. Nunn, Burt, Ross, Hughes and Valentine etc. However, the American Psychologists do not find any weight in this theory.

3. Stimulation Theory. All inner and outer stimuli that bear upon a person at one time constitute his psychological field and determine his behaviour jointly through interaction.

4. Need Theory. There are two sets of needs: *(i)* Primary or biological and *(ii)* Secondary or psychological. The more intense the need, the more the motivation.

5. Energy Theory. No energy, no motivation. According to Freud the source of physical energy is Eros and according to Jung, it is Libido. As a matter of the fact this theory offers no explanation.

6. Social Theory. According to this theory, causes of the social behaviour are to be found in the social environment, there are two streams of this theory:

(i) ***The Field Theory.*** According to this behaviour is caused by the interaction between a person and his environment.

(ii) ***Cultural Pattern.*** According to this view, an individual is cast in the mould of the culture to which he belongs. The different cultures would, therefore, produce different types of personalities.

7. Behaviour or Learning Theory. This is more elaborate than the need theory. Hull and his associates are the supporters of this theory. The theory has three main tenets:

(*a*) All motivated behaviour is based on needs and desires.

(*b*) All learning involves reward in the sense that only those responses that reduce need or drive are stamped in.

(*c*) Needs may be biological or psychological, primary or secondary.

Maslow, Tolman, Hebb and Mowrer do not share this view. They argue that all learning is not like that. Learning can be cognitive type also. It is not only the 'need reduction' but also 'avoidance of pain' that goads one to learn.

8. Physiological Theory. This theory holds that the secrets of mind are locked within the cells of the nervous system.

9. Depth Theory. Freud is the main protagonist of this theory. The spring of action is unconscious which is dark, ruthless, very powerful and illogical. Special exploratory techniques are needed to dig out the unconscious. Unconscious motives influence our conscious thought and conduct.

ADLER'S THEORY

Alfred Adler was born of a middle class family in Vienna in 1870 and died in Scotland in 1937.

Alfred Adler was one of Freud's earliest pupils but gradually he began to develop ideas which were not in conformity with those of Freud and others in Vienna Society. When these differences became very sharp he was asked to present his views to Vienna Society. Adler resigned as President of Vienna Psycho-analytic Society and a few months later, he broke off his relations with Freudian psycho-analysis. He formed his own group which came to be known as *Individual Psychology*.

Main Concepts Put Forward by Adler

Adler's view-points can be stated under a few general headlines:

1. Fictional finalism.
2. Social interest.
3. Inferiority feelings and compensation.
4. Style of life.
5. The Creative selt.
6. Striving for superiority.

1. Fictional Finalism. Shortly after Adler terminated his relations with the circle that surrounded Freud, he came under the philosophical influence of Hans Vaihinger whose book The Psychology of as if had been published in 1911.

Vaihinger propounded the idea that man lives by many fictional ideas which do not have any counterpart in reality. "All men are created equal", "honesty is the best policy" and "the end satisfies the means", enable man to deal more effectively with reality.

Adler took note of this philosophical doctrine of idealistic positivism and propounded his own design.

Adler found the idea that man is motivated more by his expectations of the future than by the experiences of the past. Neither Vaihinger nor Adler believed in predestination or fatality.

Adler like Jung identified Freud's theory with the principle of casualty and his *own* with the principle of finalism.

He states, "Individual psychology insists absolutely on the indispensability of finalism for the understanding of all psychological phenomenon. Causes, powers, instincts, impulses and the like cannot serve as explanatory principles. The final goal alone can explain man's behaviour. Experiences, traumata, sexual development mechanisms cannot yield an explanation but the perspective in which these are regarded, the individual way of seeing them, which subordinates all life to the final goal can do so."

The final goat may be a fiction, that is, an ideal which is nonetheless not attainable. This is a very real spur to man drivings and the ultimate explanation of his conduct.

Adler believed; that a normal person could free himself from the influence of these fictions and face reality when necessity demanded. This is something that the neurotic person is not capable of doing.

There were three stages in his thinking:

(i) To be aggressive.

(ii) To be powerful.

(iii) To be superior.

Adler's concept of superiority is something very analogous to Jung's concept of the self or Goldstein's principle of self actualization. It is striving for perfect completion. It is "the great upward drive". Adler writes "I began to see clearly in every psychological phenomenon the striving for superiority. It runs parrallel to physical growth and is an intrinsic necessity of life itself. It lies at the root of all solutions of life's problems and is manifested in the way in which we meet these problems. All our functions follow its direction. They strive for conquest, security, increase, either in the right or wrong direction. The impetus from minus to plus never ends. The urge from below to above never ceases. Whatever permises all our philosophers and psychologists dream of self-preservation, pleasure-principle, equalization—all these are but vague representations, attempt to express the great upward drive."

Striving for superiority or perfection is innate, that it is part of life. In fact it is life itself. It is a prepotent dynamic principle. Alder admits that the striving for superiority may manifest itself in a thousand different ways and that each person has his own concrete mode of achieving or trying to achieve perfection.

2. Social Interest. Adler was an advocate of social justice and a supporter of social democracy. He enlarged his conception of man to include the factor of *social interest*. In its ultimate sense, "Social interest is the true and inevitable compensation, for all the natural weaknesses of individual human beings."

According to Adler social interest is inborn. Man is a social creature by nature and not by habit. However, like any other aptitude, this innate predisposition does not appear spontaneously. It has to be brought to enjoyment by guidance and training. He believed in the *benefits of education*. Adler spent a great deal of his time in establishing child guidance clinics, in improving the schools and in educating the public regarding proper methods of rearing children. Social interest replaced selfish interest.

3. Inferiority Feelings and Compensations. Adler stated that inferiority feelings are not a sign of abnormality. They are the cause of all improvement in man's lot. Inferiority feelings may be exaggerated by special conditions such as *pampering* or rejecting the child which may result in abnormal manifestations, such as the development of an inferiority complex or a compensatory superiority complex. But under a normal situation the feeling of inferiority or a sense of incompleteness is the great driving force of mankind. In other words, man is pushed by the need to overcome his inferiority and pulled by the desire to be superior.

4. Style of Life. It is the most distinctive feature of his psychology and is the recurrent theme in all of Adler's writings. Style of life is the principle by which the individual personality functions. It is the whole that commands the parts. It is the principle that explains the unique-ness of a person. Everyone has a style of life but no two people develop the same style. One person tries to become superior through developing his intellect, while another strives to achieve muscular perfection. The intellectual has one style of life, the athlete another. The intellectual reads, studies and thinks. He lives a more solitary life than the active man does. He arranges the details of his existence, his daily routine, his relations to his family, friends and acquaintances, in accordance with his goal of intellectual superiority. Everything he does, he does with an eye to his ultimate goal. All of a person's behaviour springs from his style of life.

Stage Forming Style of Life. The style of life is formed very early in childhood by the age of four or five. His attitudes, feelings, apperceptions become fixed and mechanized at an early age. It becomes extremely difficult to change style of life at a later stage.

What Determines the Individual's Style of Life?

The Creative Self. In essence the doctrine of a creative self asserts that man makes his own personality. He constructs it out of the raw material of heredity and experience. He states, "Heredity only endows him with certain impressions. These abilities and impressions and the manner in which he experiences them—that is to say, the interpretations he makes of these experiences—are the bricks or in other words, his attitude in life which determines his relationship to the outside world." The creative self gives meaning to life. It creates the goal as well as the means to goal.

Concluding Observations. To conclude it may be stated that Adler gave a humanistic theory of personallty which was the antithesis of Freud's conception of man. By endowing man with altruism, humanitarianism, co-operation creativity, uniqueness and awareness, he restored to man a sense of dignity and worth that psycho-analysis had pretty largely destroyed. Adler offered a portrait of man which was more satisfying and more hopeful.

PSYCHOLOGICAL SELF-PSYCHOANALYTICAL CLASSROOM MOTIVATION

Jung (1875–1961), a fatnous Western psychoanalyst who transcends both Freud and Adler and goes deep into the human psyche revealing some of its mysteries. He seems to be more of a philosopher and less of a psychologist. Starting as one of Freud's associates he started differing with him and by 1912 he developed serious ideological and personal differences and founded his own school named Analytical Psychology.

Jung considers Life itself as an 'ever living fire' and self-realization as the aim of life, which is far beyond satisfaction of sex and seeking of an easy-chair high prestigeous job. There are, of course, persons who are mad after these two things, but this does not minimise the importance of 'self-realization'. We may brush aside such differences as individual differences.

Jung does not question the importance of sex and status but he goes beyond these. He often appears to attack Freudian Concept of Sexuality. He stresses certain clear-cut aim of life before the individual. He states, "I have observed that a directed life is in general better, richer and healthier than an aimless one and that it is better to go forward with the stream of life than backward against it."

We are living in an age of anxiety and tension and lawlessness in the outer world. Western man expects that psychology should be able to assist him in understanding the incomprehensible factors of life. Jung thinks that there cannot be law and order in this outer world so long as man does not feel at home in the inner world, in his own soul. Jung is very near the living process of the soul. He wants to lead people to responsible attitude towards life and a responsible way of living, suited to the peculiarities of each man. Jung is optimistic about human nature.

The Structure of Self. Jung does not deny the outer world but he gives importance to the inner world and the deep-seated unconscious. He gives a complete picture of the "Self" with its various aspects in the outer as well as the inner world. Jung thinks that the term 'self' stands for the integration of the various aspects of the personality, conscious and unconscious. Development of personality takes place when an individual chooses his way and separates himself from the crowd. His concepts about the different aspects to the self are woven into one integrated whole. He accepts conscious as well as unconscious. But his concept of unconscious is entirely different from that of Freud.

He thinks that the unconscious is the matrix out of which the conscious has been evolved. Unconscious is not evolved out of the conscious but the reverse. Consciousness is the product of growth, whereas unconsciousness even existed before the intellectual self began to function. Our consciousness arises from unknown depths. Jung's concept of the unconscious is that of something vaster and more comprehensive. According to Jung unconscious was just not the dumping ground of the repressed desires from the conscious (as Freud thought).

Jung admitted Freud's explanation of the unconscious but felt that it was only a part of the vaster unconsciousness, which is shared by an individual with all other people in all places and at all times.

Jung has classified unconscious, in two types:

(1) Racial consciousness.

(2) Personal unconsciousness.

Racial unconscious consists of primitive beliefs and modes of response handed down over centuries. We call these ancient was of thinking as 'archetypes.'

The unnoticed, forgotten or repressed incidents of each person's life form personal unconscious. Here, he agrees with Freud.

The unconscious of an individual is opposite to his conscious mind and is complementary to it. The unconscious tries to correct and moderate the conscious.

We can compare the picture of the mind given by Jung with compensatory aspect of a photograph of a range of mountains mirrored in a lake below. Just as each mountain has its equal and opposite reflection, in the same way each mental attribute has its equal and opposite complement. To Jung, therefore, a man is both better and worse than he knows. The coward is a brave man and the brave is a coward in his unconscious; the romantic is realist and *vice versa*. Different layers of unconscious according to Jung are as follows:

(a) Animal life.

(b) Primitive ancestors.

(c) Ethnological groups (races).

(d) Nations.

(e) Family.

He further observes that conscious mind permit itself to be trained like a parrot, but the unconscious does not. Our uncanscious contains the experience of the whole race. Jung states, "Bady is a sort of museum of its phylogenetic history, so is the mind".

According to Jung the unconscious is more than just containing personal memories. The collective unconsciousness represents a sort of social memory. He also points out that certain identical customs are found in widely scattered parts of the earth. It has also been noted that Siblings' dreams are very similar even though they may be hundreds of miles apart in different milieus.

There is only racial unconsciousness at birth. It has got three things at that time:

(i) Individual instinctive reactions.

(ii) Ancestoral behaviaur.

(iii) Ancestoral interpretations, *i.e.*, belief in ghosts, fairies etc. The unconscious forgets nothing. It is mnemic in nature.

Jung has archetypes, symbols:, Libido, Shadow, Persona and The Soul Images while explaining the concept of unconsciousness. These concepts are explained here:

Archetypes. Ancient ways of thinking and symbols or figures which explain life are called archtypes. Symbols are generally based upon experience of the individual but the experience of the individual is the experience of the human race, the collective unconsciousness. The collective unconsciousness is the "all controlling deposit of ancestral experience from untold millions of years" to which is added by each century small amount of variation and differentiation. These experiences are archetypes. Their sources are to be found in profound religious thought, in enduring literature and art which draw their universal inspiration from the depths of human psyche.

The archetypes are eternally inherited identical forms and ideas. The archetypes may appear as persons. They may appear as demons or as fabulous animals or natural forces or objects or geometrical forms and numbers, etc. Jung says, "Archetypes resemble the beds of river. An archetype is something like an old water-course along which the water of life flowed for a time, digging a deep channel for itself. The longer it flowed the deeper the channel and the more likely it is that sooner or later the water will return."

Archetypes were and still are psychic forces which must be taken seriously. The archetype is a psychic organ present in all of us. They are complexes of experience that come upon us fatefully. They affect our personal life. We can never legitimately ignore our archetypal foundations.

Shadow. Shadow represents the unconscious and is closely related to us as the dark reflection of our conscious efforts. It may become monstrous and may actually overwhelm the ego-conscious. In one sense the shadow is a highly personal creation. In another sense it reflects the universals of human experience and is 'personal' only in a very limited manner.

Symbols. They are of greater psychological value as they are necessary openings to understand 'human experience'. They serve to reveal the depths of the soul. They are created by deep unconscious and are found all over the world. They appear in dreams, art productions, visions and fantasies. These symbols lead to the expression of archetypal, thinking.

Libido. Jung while retaining the Freudian term 'libido' gives it a different meaning. He means by the term a life energy underlying all natural phenomena, including the human psyche. He believes that sexuality is one manifestation of libido but not all. Libido is essentially creative and it takes many forms. For Jung libido is life itself. It may follow different channels, with the individual gaining maturity, Sexual is perhaps the most important among these but not the only one. He took it as the life urge. Jung suggests three stages of development. *(1)* Pre-sexual, upto three, four years the infant has epoch of sucking nutrition only, *i.e.,* mouth zone. *(2)* Sharp changes take place when the child leaves the mouth zone and the individual develops and learns to turn from parents to the real world around him. *(3)* Libido gets direction in two ways—Introversion and Extroversion.

Jung's concept of the ego-conscious is similar to that of Freud's.

Persona. Persona is adaptively organized image of the self which all of us form in the course of living. It is forpted consciously as we develop our goals and ideals. Extroverts can hange persona as they change dress. It may differ at different times, places and persons, identification with persona is full of dangers, it represents the face show to the world.

The Soul Images. The soul image (Anima-Animus) lies deeper in racial unconscious. Every woman has her animus, the male component. The anima includes archetypal images of woman, the eternal mother, the virgin, the Harlot, the witch, the temptress, the spiritual guide like Dante's Beatrice.

Psychological Types. Jung believes that libido takes two directions—Extroversion and Introversion. Both are necessary for complete living. Freud's stress on sexual relations is extroversive and Adler's insistence on status seeking is introversive. Jung considers both these right but he goes beyond them and states that these attitudes complement each other. Even within himself the individual has the counterpart. The extrovert wants to a just himself to the reality of his environment and the introvert on the other hand seeks to adjust himself to the world which lies within himself. He thinks that the contents of the unconscious possess a greater reality than the things of the outer world.

Each type performs four functions—sensation, intuition, feeling and thinking. All the four are essential for ideal living and should work in harmony. Sensation concerns the present intuition with the potential—what the living will be or has been, feeling gives us a sense of values and thinking tells us what the things is in an abstract manner.

Thinking coupled with intuition may take us away from reality and coupled with sensation, it may bring us down to reality.

With each of these functions the same person may have one dominant general attitude in the conscious region of the mind and the other in the unconscious. According to these functions total personality types fall in eight categories. All persons may be ranged in a continuously graded series ranging from extreme introversion to extreme extroversion.

❋❋❋

8

Personality and Character Assessment

Personality is the most abstract word used in English language. Its connotative significance is very broad but denotative significance, negligible. Scarcely any word is more versatile. There is no single definition, which may be regarded as correct. Some of the definitions are psychological some are not. The first task is to distinguish between them and the second task is to select from among available psychological definitions one that best fits the phenomenon one wants to assess.

The word personality has been derived from the word 'persona' which means theatrical mask. 'Persona' has four distinct meanings:

(a) What one appears to others and not what one really is.

(b) The part one plays in life.

(c) The sum total of qualities.

(d) Distinction and dignity as in style of writing.

According to the first meaning personality may be thought as external appearance and not the true self. The second meaning of persona regards personality as a role, which the player assumes. The third involves distinctive personal qualities in the personality and the last derivative has significance of prestige and dignity.

Philosophers, psychologists, sociologists and theologists have defined personality in their own way. The church fathers used the term persona to designate the members of the Trinity and what was once thought of assumed character came to have the opposite significance of inner nature or substance or even essence. The philosophers regarded personality as sum total of distinctive qualities of rationality, self-consciousness, thinking and hence, an intelligent person gifted with powers of understanding and reasoning was thought to be imbued with unique personality. The sociologists regarded personality as the integration of all those traits which determine the role and status of the person in society and defined it as social effectiveness.

Psychological Definitions

Psychologists from the classical times regarded 'persona' as the assemblage of personal qualities and even now psychologists define personality as :

(*a*) The sum total of all the biological innate dispositions, impulses, tendencies, aptitudes and interests of the individual and acquired dispositions and tendencies got from experience.

(*b*) The entire organization of a human being at any stage of his development.

(*c*) The integration of patterns, which give a peculiar individual trend to the behaviour of the organism.

(*d*) The organised system, the functioning whole or unity of habits, dispositions, sentiments that characterise one member of a group as being quite different from any other member of the same group.

Gardener Murphy's View. Gardener Murphy has pointed out that the word personality has been used by the psychologists in three different senses all of which are closely related to each other. Personality has been defined as unique individuality or the inner organization of traits and tendencies or the way of behaving. When we speak of a man's personality we keep in view the external pattern of qualities which make him different from other people or we try to look at him from within and regard his personality as a complex organization of psychological factors or we may mean the way he deals with his circumstances, the way he reacts to persons and things and innumerable stimuli of his environment; *for example,* he may react to them calmly or emotionally, confidently or intelligently, consistently or inconsistently.

All these meanings or definitions are correct and some are not even sharply differentiable and hence, not inconsistent. However, we have to choose from among these definitions or meanings the one which suits our purpose best which is assessment of personality. In simpler words, we can say that the child's personality which we want to assess, is the sum total of his ways of dealing with his circumstances. By sum total we mean not the collection but the organization and interrelation of his modes and methods of dealing with circumstances into a systematic pattern. This conception of personality then regards it as the total pattern of his ways of dealing with the challenges and problems of life.

Allport's View. In more difficult and psychological terms, Allport defines personality as the dynamic organization within the individual of the psychological systems that determine his unique adjustment to his environment.

The system that constitute personality include habits, physique, temperament, skills, interests, fears, hopes, likes and dislikes, looks, feeling, knowledge, specific and general attitudes, sentiments, dispositions and traits and so on. The organization of these system is in body and mind of the person concerned; hence, these systems are psychophysical. The organization of these psychophysical systems is dynamic because it is not only constantly evolving and changing itself, but it is self-regulating too. Endowed with these dynamic psychophysical systems the individual adapts himself actively or passively, intelligently or confidently, consistently or inconsistently, completely in his own unique way. Hence, the best definition of personality would be the dynamic organization within the individual of the psychophysical systems that determine his unique adjustment to his environment. In none of us is personality static but is subject to growth influences, hereditary potential, environmental factors and personal reactions.

Characteristics of a Good Personality

It will not be out of point if we throw some light on what good personality is. The environment of an individual offers him problems and challenges of life. The more successfully a man deals with these challenges and problems of life with the help of higher mental processes of reasoning, drawing inferences, analysing and understanding or with what we call creativeness or inventiveness, the greater claim he has for possessing a good personality. The man of good personality may also be regarded as one who has a varied range of living interests and vital concerns. Allport says that a man of good personality has strong participating interests in the following six value spheres—the political, the economic, the recreational, the religious, the cultural, the scientific and the domestic. Mental hygienists lay greater stress on wholesome personalities. By a wholesome personality we mean :

(a) A willingness to recognise one's own limitations, abilities, failures and attainments.

(b) A good and satisfactory social adjustment which means on the one hand satisfaction of our needs of social recognition and respect and on the other achievement of a social status, a vocation, a place in the society.

(c) Consistent and integrated personality traits free from conflicts, complexes, frustrations and contradictions.

(d) Working adherence to social ideals.

When we say that personality is the whole man we mean to say that his personality involves what he is, what he can be and what he hopes to do.

Kretschmer's Classification. People may be either athletic, pyknic, leptosome or dysplastic.

Jung's Classification. Jung has classified people into two types-Extrovert and Introvert. The extrovert is more social, the introvert is more self-centred.

Spranger's Classification. Spranger typifies personalities as follows-theoretical, political, economic, religious, social, aesthetic.

Factors Determining Personality

Personality is the total configuration of individual characteristics and modes of behaviour that shape one's adjustment to his environment. As personality is a complex concept we have an interaction of a number of factors that go to shape the personality of an individual. Some of the general factors are :

(1) Genetic Factors. The effect of genetic factors on personality is not direct as there is little evidence that there is a causal relationship between inherited factors and personality traits, such as co-operativeness, aggressiveness and honesty. Hereditary factors exert their effect upon the organism through the constitution make-up. The organism responds to the external world with his psychological mechanisms which the hereditary factors control. Heredity affects the endocrine gland structure which in turn affects the behavioural traits of an individual.

Today it is believed that genetic factors produce disposition or potentials for development rather than directly producing bodily or behavioural traits (Lazarus, 1961). It is also believed that heredity sets the limits beyond which individual may not develop. The importance of heredity seems to be small in mental development and great in physical characteristics. The interaction of heredity and environment can be illustrated by the famous study of Jukes family or of the descendants of the Kalukak. Of 1200 descendants in Jukes family most were criminals, paupers and dependents. These people, who had poor heredity, lived in poor cultural environment; similarly among the 480 descendants of Mastin Kallik 143 were feebleminded and many were sexually immoral, intemperate, delinqent or criminal. Thus, heredity and environment both play an important role in the shaping of personality characteristics. Similarly, the descendants of Jonathan Edward and Sarah Pierpont were college presidents, professors, authors, lawyers, judges and statesmen.

Hereditary plays an important role in endocrine structure and its functioning. Many desirable and undesirable qualities of personality can be attributed to glandular balance. The endocrines such as thyroid, pituitary, pancreas, gonads and adrenals pour hormones directly into blood stream. If these sections are not available in delicately balanced proportion, there are marked deviations in intelligence, emotional control, growth process, appearance and behaviour. A teacher should recognise that glands play an important part in human behaviour and he should not condemn human behaviour. *For example,* laziness in children may be due to hyperactivity of thyroid which may be cured through thyroxin and the hyperactivity of thyroid may force a student to seem unable to stay in his place, may impel him to demand attention by raising his hand to answer the teacher's question. The teacher, who recognises that it is because of the hyperactivity of the thyroid gland, would tolerate such a deviant behaviour.

Some endocrine imbalance results in a change in blood sugar and blood calcium. In such cases pupils become inattentive, antisocial, often negativistic and have other physical symptoms. Other endocrine glandular imbalances have effects on behaviour but that effect cannot be so readily identified as we have in the case of thyroid gland. *For example,* the psychological effect of the gonads, pancreas, adrenals is variable because it has been observed that the same defect may cause one child to be meek and retiring but make another aggressive. A child or a youth suffering from a glandular imbalance is more susceptible to the type of treatment at the school and in the family. Teachers and parents should be careful while dealing with him; because it is not the hereditary factor alone, but the environment also that determines the nature of a pupil's behaviour.

(2) Effect of Body-build on Personality. The laymen believes that the big muscular boy is dominant and aggressive, that the fat girl is jolly and easy going and that the lean and thin person is serious and tense. Some psychologists have tried to relate body structure to personality traits. Kretschmer hypothesised that the pyknic body type (thickset, round person) is gay, good-humoured and happy; that the asthenic body type (learn, underdeveloped person) is polite, sensitive and cold and that the athletic strong boy has a personality between the other two types. The dysplastic, who has a thickset upper body and either lean and slight or muscular lower body, is a mixed type.

Body build has a relationship with personality trait, but it should be remembered that in most cases the relationship between physique and behaviour is a result of social behaviour; *for example,* a child who is suffering from some physical handicap is rubuffed by his peers and is considered less worthy than the normal one. No direct causal relationship can be assumed to exist between body structure and personality traits.

The converse of the above assumption holds good in most cases. When the physical handicap is in socially approved directions, *for example,* when the girl is very attractive or the boy is very strong, she or he is found above average in social and personal adjustment. One cannot definitely predict personality characteristics by considering the body-build or a physical handicap; *for example,* the fat boy may be good-humoured not because he is fat, but because he cannot fight. The body-build theory of Kretschmer is not tenable also because it tends to place human-beings into categories and ignores the moulding influence of environment.

There is another theory by Sheldon which provides a continuous variation in type. A given individual may possess one or more characteristics of each of the following types :

(a) Endomorph (a person who wants physical gratification, a pleasure lover).

(b) Mesomorph (likes physical exertion, is energetic and loves competition).

(c) Ectomorph (self-sufficient, enjoys intellectual pursuits, a sensitive nature).

The endomorph is a heavy-set person; the ectomorph is thin and the mesomorph is muscular. This theory is also not tenable because an endomorph may be scholarly and introverted and hence, possesses the traits of an ectomorph. Body-build and personality traits are not connected in a cause-and-effect relationship. Teachers should not, therefore, infer personality traits on the basis of superficial observations of physical characteristics.

(3) Environmental Factors. The environmental factors that play a significant role in shaping the personality of an individual are his home, his socio-economic status, community's educational institutions, government, community, religious institutions, local customs and mores. The environment may act against the influence exerted by heredity. Children may develop personality characteristics like or diametrically opposite to those of their parents.

Home is the most important environmental factor that shapes the personality traits of children. The effect of infantile experiences have been studied by taking infants into orphanages, hospitals and foster homes and by comparing their behaviour with the behaviour of children cared for by their parents in their own homes. It is alleged that adverse infantile experiences lead to progressive intellectual decline, poor social adjustment and other behavioural problems. One of the defects of such studies is that the children taken to institutions do not form a representative sample of the general population because it is true of those placed in hospitals. Institutionalised children have been found superior to those reared in their homes.

The following relationships have been recurringly noticed in replicated research studies (Brown, 1965) :

(i) Parental warmth and withdrawal of love is related to strength of conscience in children.

(ii) Withholding or bestowal of tangible rewards and physical punishment on the part of parents leads to aggressiveness of behaviour and poorly developed conscience.

(iii) Parents' stress on early independence and strong achievement motivations in the child are positively correlated.

Socio-economic status of children and youths affects their personality traits through different social goals, ideals and attitudes. Different socio-economic conditions in a community or in a home produce their effect on attitudes of persons regarding problems of morality and religion. When the child enters the school, he is given the same educational opportunities and the same experience, but he chooses to learn what his socio-economic status decides. All pupils get the same material from the teacher, but they do not learn the same things. What they learn is certainly influenced by the socioe-conomic status he belongs to. The social class has had determined the path of life of child much before he enters the school. The school can only encourage the pupils to rise above that status and thus, stimulate them to overcome the environmental handicaps.

The School can also improve the home conditions by giving parents proper guidance as regards in improvement in nutritional standards and beautification of homes and other sanitary conditions. It can modify curriculum to fit the needs of children. It can arrange for kindly-disposed and realistic teachers. It can do a great deal in reducing delinquency, feeble-mindedness, dependency. "We do not really know how much the mind can achieve until-consistent effort is made to develop its potential."

Society. The neighbourhood, sub-culture and general culture have also a potent influence upon what a child will be. The patterns of behaviour and standards of one's peer group and other social pressures exert a great influence on one's behaviour.

Thus, we see that the personality development is a resultant of two main factors :

(i) What we are (heritage).

(ii) What we have (environment).

But there is also another factor which should not be lost sight of. It is the individual's response. Two persons perceive the same stimuli but they react to them in quite different ways. One's response is an outcome of one's potential (heritage) as acted upon by environment. Thus, one's personality is the resultant of three forces; what one is, what one has and what one does.

Methods of Assessment of Personality

The methods of assessing personality are as varied and complex as the phenomena assessed. The concept of personality is so abstract and the term is so versatile that no single method may be regarded as a legitimate one. The most legitimate method of assessing personality is observation of a datum coupled with the interpretation of its significance. However, for studying personality the following list of psychological methods are suggested.

Subjective Methods

1. Observation. The psychologist observes the behaviour of the child for many days. Generally he makes a note of particular personality traits possessed by the child and gives his own assessment.

2. Case Study. The psychologist collects every kind of information and records it. All the traits of personality are viewed. All the sources are exploited. From all the records and intelligence, aptitudes, achievement test scores a dependable opinion is formed.

3. Inventory Method. Inventory is again a set of questions whose answers are to be given as yes or no. Inventories are valuable for measuring such traits as sociability, emotionability, suggestibility, honesty, truthfulness, interest etc.

4. Interview. The interviewer asks a number of direct questions to the child and observes his apparent trait. He probes into his interior self through various questions. Interview being completely subjective is not an objective method of personality testing. Some children who are upto the mark in a particular trait become nervous and suppress some valuable information. The assessment then becomes quite faulty.

5. Autobiography. The child is asked to write down his own autobiography. Autobiographies reveal the writers personality characteristics well.

Objective Methods

1. Rating Scales. A particular personality trait is rated on 3, 5, or 7 point scales which indicate the amount of the trait possessed by the individual.

2. Questionnaire. Questionnaire like inventory is a set of standardised questions the answers to which reveal a person's traits.

Projective Techniques

Projection means ascribing one's own feelings and attitudes to people or objects outside oneself. When we look at an unstructured cloud shape, we see some strange figures having feelings as nearly as our own and we project into that cloud shape our own tensions, anxieties, worries or attitudes. It is a human trait that we read into an unstructured material-an event, a cartoon, a picture, a cloud shape-meanings that can come only from our own personality. Everyone of us interprets an unstructured inkblot, *for example,* in his own way and tells what he sees in it. It is assumed that we project our own personality pattern into that picture or inkblot or situation. The necessary and sufficient condition for that assumption is the unstructuredness of the object or material. We may also project our personality pattern into still less prestructured objects by drawing, carving, or by structuring unshaped material.

Assumption. Such projective techniques are believed to have the signal advantage of leading a person to reveal more of his personality pattern that he would if faced with a set of direct questions. Many tools have been evolved for assessing personality in this way. Two of these are T.A.T. and Rorschach's Inkblot Test, both of which stimulate the testee to project his innermost feelings and dynamisms. The subject's repertory of tensions, anxieties and attitudes is analysed in terms of certain criteria or a frame of reference. The interpretation of subject's responses is made according to that frame of reference or criteria. Clinicians believe that if we want to make a probe into the personality disorder of a man we must first gain insight into the complex causes conducive to such a disorder. They claim that while reacting to a given stimulus the individual discloses at least a partial explanation of his behaviour. While projecting his innermost personality pattern, the subject forgets for the time-being the conventional pose that he generally assumes and thus, reveals his true self. It is assumed that the inner self of some kind exists which can be measured and analysed. It is also maintained that the abnormal as well as the adjusted persons conceal many of their basic tendencies and desires in conforming to the moods of the society. These they reveal in their responses to projective tests. The vague and unstructured material frees them from social conventions and from reality so that they can feel themselves.

Rorschach Inkblot Test. Hermann Rorschach, a Swedish Psychiatrist prepared to inkblots: 5 Black; 2 Black and Red; 3 in different colours and standardised them. They present an unstructured situation before the subject whose personality is tested. The Rorschach Test purports to measure the hopes, anxieties and tendencies of the individual, all of which are known as motivation and affect. If the assumptions male for projective techniques are correct, the Inkblot test can be regarded as a valid instrument for the measurement of personality. One of the assumptions of the projective techniques is that the inner dynamically related tendencies and attitudes are discernible through responses to vague situations and can be measured, then the following claims can also be made and be justified for the Inkblot Test :

(*a*) Specific mental processes can be isolated.

(*b*) The test can, therefore, be said to differentiate between Erlebruis type and the experience type in terms of their respective roles which the introvertive and extratensive trends play in the personality structure. The personality of an individual is affected and structured to the extent to which such factors motivate or repress the individual who is in search of personal and social adjustment.

(*c*) These processes can be understood in terms of their interplay and the degree of balance and equilibrium attained.

When the individual is asked to relate what he sees in the inkblots one by one, he offers certain responses. Each response is related either to location or determining quality or content. It may be based on the whole or on a detail of the smear (location) or it may be determined by colour, shading or movement (quality) or it be connected with human figures, animals, objects, landscapes, maps, anatomical details (content).

His scores are added or quantified. Other scores are also noted down; *for example,* Reaction Time to each blot.

The interpretation of the scores is a complicated task because formal experimental proofs alone cannot determine the personality structure of the individual who is given the test. The skilled interpreter has to base his conclusions on logic and psychologic experience. *For example,* a person, who gives more of whole responses, is theoretical and not practical and the one who gives more movement responses is introverted or attached to fantasy.

The special features of the responses given in the Inkblot test are :

1. Certain responses to ink blots are consistently related to personality disorders.
2. The quantification of scores or weighting or responses can help up in the identification of certain categories of personality disorders, *i.e.,* psychoneuroses, psychopathic personalities, brain injuries and schizophrenia. These categories are arranged on a scale determined by the Scores.

Caution. A word of caution may be given. Other methods of assessing personality should also be used by those who use this technique, for we cannot exclusively rely upon the conclusions made on the basis of this test. People may have advanced one-sided and exaggerated claims for diagnostic utility of this test. Bust the wise investigator will not place his reliance on this test alone.

Thematic Apperception Test. The Thematic Apperception Test contains 10 to 20 unstructured or vague pictures, the characters of which are drawn with vague lineaments. Most of these are human figures sufficiently ambiguous to permit a variety of interpretations by the persons tested.

The test assumes that while the subject sees these vague pictures, he will reveal his inner action-picture in his responses.

It is also assumed that the subject will project his feelings, his likes and dislikes, his anxieties and hostilities into those shapes. Generally, the subject identifies himself with this or that character and in describing what is happening or what led up to this or what will be the outcome, the subject describes his own fantasies, wishes, conflicts, desires. Thus, the story he depicts becomes entirely autobiographical.

The test cannot be scored quantitatively. The investigator records the themes that come again and again more than one picture and analyses: *(i)* the emotional tone of the story *(ii)* the needs and desires of the subject *(iii)* his conflicts and frustrations. The test affords a conception of the subject's personality, which is rich and suggestive though not quantitative and precise.

Personality Inventories. A personality inventory is a detailed list of questions the responses to which are intended to measure a certain personality trait. Maudslay's Personality Inventory is an example. It gives a standardised measurement of personality. Most personality inventories lack comprehensiveness and are not valid. They are valid only for special purposes. They do not heir one who desires easy and accurate evaluation of personality characteristics. The use of personality inventories and their popularity have been due to the people's psychometric innocence and their mistaken faith in statistical measures. The use of personality inventories in all cases is also not desirable because they are static instruments and cannot reliably measure a personality characteristic which is always dynamic. Their use is also not recommended on the ground that one's response to an item depends on how he interprets the question. The instrument is subject to dodging. If the correct response to an item is required, an establishment of rapport with the examiner is necessary, otherwise the testee is likely to ask, "What business is it of his to ask me whether I am moody or not?" One's mood differs from time to time. A happy or unhappy experience, a short time before taking the inventory, might change the response and lead to a wrong final score. The man, who uses a personality inventory, or who is not familiar with its limitations may, reach wrong conclusions. So, it is necessary that the results of a personality inventory should be viewed in the light of the results of other measurements of personality.

A few items to illustrate a personality inventory are given below :

1. Do you sometimes envy the happiness that others seem to enjoy?
2. Do you get angry easily?
3. Do you have a strong desire to run away from home?
4. Do you have spells of dizziness?

Rating Scales. Rating means comparing one individual with others in the group. The rater gives his subjective judgment of the extent to which a person has a particular trait in him. Rating may be done by the individual himself or by some other person. A rating scale consists of questions, such as "Is he liked by his classmates ?" The degree to which a boy is liked by his classmates is indicated by numbers 5 to 1 or 7 to 1 and descriptive phrases are assigned to each number; *for example,* 5 may be allotted to 'exceedingly liked' and 1 to 'very much hated'. As there is a halo effect that influences the rating done by one person, the ratings of several other persons are averaged.

The self-rating scales are used to compare the picture of personality as perceived by others with the picture of personality as perceived by oneself. 'Guess who' is also a sort of rating scale in which peers rate each other. They are required to tell the name of the person who laughs rarely or talk most or is friendly and so on. Such a scale may help the teacher in modifying his judgment about a pupil by comparing the picture given by his peers and that formed by the teacher himself.

CHARACTER

Some people think of character as a moral quality; others think of character as a bundle of habits; still other thinks that character is primarily composed of ethical concepts. Many others regard character a reflection of emotional adjustment. Character is not really a sum of separate habits or qualities. Character is embedded in the total structure of personality. Therefore, character is closely related to personality and to gain or to understand the nature of character, we will have to examine the very structure of personality. But, as an examination of personality structure would take greater space, we are inclined to describe character in short as follows:

We define character in terms of the choices the person makes when his actions affect the welfare of others. It expresses the person's beliefs, his social relationships and his intellectual development. A person of good character makes a choice of such acts that promote the welfare of others. These acts are consistent with his own goals.

This description of character excludes the temperamental qualites because they are not specially pertinent to the welfare of the community. We cannot judge character from an act alone. We must consider the basis on which it is made. We cannot judge the goodness of an act if we do not know what it means to an individual.

Character development can be achieved at following five levels: *(a)* Amoral; *(b)* self-centred, expedient; *(c)* Conforming, conventional; *(d)* irrational, conscientious; *(e)* Rational, altruistic.

Levels of Development of Character

The child starts at the amoral level. He does not know what effect his act would have on others. Such an amoral behaviour is rare among school-age children. If a person is amoral it shows that he has not succeeded in socialization.

The expedient person shows a far more controlled behaviour. Though he wants the satisfaction of his own wishes only, yet he knows the importance of considering other's reactions also. The, key to his low level morality is his "me first" attitude. The expedient person disregards the welfare of others which it is to his advantage to do so.

The conformer follows the standard by his group rather than any moral principle. He will follow his impulses if they do not run counter to the group norm. When the group norm changes, he changes his standard. Character should develop beyond this conventional level. The person should know why a certain group norm is desirable to conform and when to reject the norm as a model.

Character develops at the irrational conscientious level also, when the person acts in the light of values held emotionally rather than rationally; *for example,* the person, who has been subjected to severe punishment for cheating in the examination, will not try to cheat even when there is no possibility of his being detected and punished. Even if in the whole examination centre mass copying goes on he will not violate his own moral integrity. This person subject to severe punishment on some occasions develops a conflict that checks his rational discrimination.

The rational person adopts standards of conduct because he thinks that they have beneficial effects. He considers both his own good and that of others. Character is a unified structure and not a collection of isolated good habits.

Role of the School in Character Development

A person adopts certain standards according to his attitude towards authority, peers and himself. From infancy onwards he has a regular process of development of personality. The formation of personality begins in infancy and it develops in the home as he proceeds in age. The home lays a good or bad foundation for school's work. The school should not think that it is omnipotent. The emotional influences of the early years may produce in the child fears and enthusiasms which the school cannot change or modify.

The school can also not outweigh all other influences on the pupil. The pupil is far more deeply influenced by his associates and the parents than by the average teacher. Nevertheless, every pupil has a potential for becoming a man of character and the teacher should try his best to develop his character.

If the school wants that the character of children should be formulated along correct lines, it should provide emotional security. Then it may instill in them a desire to be good. The school should teach the pupils to solve their ethical problems intelligently. The experiences that they get in solving their conflicts may be changed into generalizations.

Extracurricular activities and student government may be organised on sound lines because they introduce the pupils to problems which curricular programme can not.

Difference between Character and Personality

Character is related to those socially relevant activities which have particular moral and ethical implications. The rules of conduct or moral principles which society approves as good are followed by a man of character. *For example,* a man, who speaks the truth, is honest in his dealings with others. He follows certain moral principles and we call him a man of character. When we deal with activities commonly labelled good or bad and when these activities are considered with special reference to their moral implications we speak of character rather than of personality.

Character and personality traits cannot be sharply differentiated. Personality is broader of the two concepts. Personality is character evaluated and character is personality devaluated. Since ethical evaluations differ from culture to culture, from one generation to another, the distinction between character and personality is relative and constantly shifting. Even then a few traits may be regarded as character or personality traits exclusively. *For example,* lying, cheating and stealing are dominantly character traits whereas dominance and submission, introversion and extroversion are principally personality characteristics.

The development of personality and character are essentially alike. They differ only in their evaluative components. Consequently, sometimes one term is used and sometimes the other, depending upon the relative importance of the evaluative component. However, what is said about one largely applies to the other.

Permanence of Character Traits

Everyone recognises that the consistencies postulated in the concept of character and personality are relative matters. The socially relevant habits, beliefs, attitudes, ideals and values and self-concepts which constitute the major parts of character and personality are products of learning. The acquisition, extinction, modification, organization and reorganization of these components never cease. Character and personality both are stable and unstable, constant and inconstant, consistent and inconsistent. Personality and character are relatively permanent behavioural characteristics of the individual. William James (1907) and Sigmund Freud (1933) have emphasized the permanence of character traits. James said that "by the age of 30 character has set like plaster and will never soften again." Freud agreed with James but set the age for the fixation of one's basic character traits much earlier. He believed that one's fundamental character traits are irrevocably fixed in early childhood.

Consistency of Moral Character. When we speak of some person as having moral character, we say something meaningful concerning him. We feel that he has some specific traits, a representative sample of his behaviour and assume that we can predict from these his future behaviour.

Cross-sectional studies show little consistency in the manifestation of character traits from one situation to another. There is positive but very low correlation between moral conduct in one situation and moral conduct in another situation. The young child's development in the areas of moral conduct is pre-dominantly uneven. Character trait consistency is a relative thing and has been shown to be related to several variables. Consistency of attitudes and moral conduct increases with age. There is a tendency for high consistency to go with high honesty and for low consistency to be associated with dishonesty hold Hartshorne and May. When children have been classified into two classes it is found that both groups become more consistent in moral behaviour as they get older but that the higher class children become more honest while the lower social class children become more dishonest.

Character Formation

William James in his famous essay on habit expressed the idea that character consists of specific habits and to develop character frequent repetitive practice for fixing habits is the sure method. Suppose, a person is to form habit of generosity. He may be given chances to be generous in charity drives, but how can he be taught to become generous. Suppose the teacher asks him to get money from his father; the child gets the money if he can, so that he may not be punished in the school. If he is to be criticised for not giving publicly he will be giving publicly. If he is not criticised, he will not be generous.

Practice may not develop understanding of values, because it may not produce satisfaction from good conduct and may not transfer. Schools have employed repetitive practice to teach patriotism or responsibilities in carrying our duties. For this purpose, in every morning assembly, National Anthem is sung; National Flag is hoisted over the school building on the Independence and the, Republic Day. In teaching History and Social studies lives of great men are very much emphasized. Similar procedures are employed to teach responsibilities in carrying out duties but the habits of conduct do not become fixed in the character of the children. An act loses its point, when the promise of reward or threat of punishment is removed. Those boys, who are conscientious, are more likely to retain the behaviour, not because of extensive practice but because they are willing to make this habit of conduct a part of their life. Thoughtless habit can never successfully be formed.

Deriving principles of good conduct from the pupil's activities is helpful in forming conscious ideas. Talking and reading about good behaviour has been relied on to build character by both religious and secular educators. Many books contain stories which are included there to inspire high ideals. The curriculum builders have always hoped that Literature, History and other subjects would contribute to character. These hopes are not realized. Some of the earliest experiments on character education confirmed the significance of verbal concepts. When boys are given many opportunities to discuss moral principles along with their activities and when the teacher makes a continual effort to classify and emphasize a code of good conduct students try to learn it. Therefore, deriving principles of good conduct from the pupil's activity is helpful in development of character.

Verbal education runs a great risk of being verbalistic education. Suppose, *for example,* the teacher wants to form good habits or develop character traits through giving a lecture in the classroom or a talk in the morning assembly. His students may learn the words denoting moral characteristics, but there is no Certainty that they will learn the sense or accept the principles as their own. If the students are required to adopt moral codes of conduct, they should try the principles in real situations and the school should provide opportunities for such application so that they may understand how these moral principles work in real situation.

Verbal principles have only a surface influence, unless the person incorporates them into his ideals. Students know that cheating is sinful and it is also harmful to them but it is strange to find that when a good chance is given they are found cheating. It is also strange to find that delinquents have more moral and religious knowledge than non-delinquents according to E.R. Bartlett and Harris. Researchcs by Corey (1937) indicate that unless verbal principles are incorporated into the ideals of a person they may remain superficial.

Since conduct is rooted in personality, the starting point for character education is the reduction of tensions. If the school wants to promote good conduct, it should do two things :

(*a*) Reduction of frustration.

(*b*) Helping students in accomplishing the developmental tasks.

Every activity that helps the student to accomplish the developmental task and gain self-acceptance, lays a foundation for rational, altruistic conduct. The rational person adopts standards of conduct that he perceives to have beneficial, effects considering both his own welfare and of others. Such a person is said to possess rational, altruistic conduct.

Schools should try not to create thwartings in any way. Every child should be given an environment, in which he gets a sense of success. Instructional programmes should be made more meaningful. The individual weaknesses should be diagnosed. Handicaps should be removed. In this way frustrations and tensions can be reduced and once the child's tensions are reduced, he tends to show good conduct.

The school should help the pupil to see authority as reasonable rather than threatening. If the child is not forced to defy the orders or obey them through fear, he will certainly develop good qualities. The school that promotes emotional security and understanding of values contributes a great deal to character development.

9

Delinquency, Exceptional and Backward Children

Juvenile delinquency is one of the most serious problems of our times. Everybody has his or her idea of what delinquency is but many of these concerns do not take us anywhere. As such, we need a clear-cut definition.

1. A juvenile delinquent is one who is a minor with major problems.
2. The best definition, "Delinquency is anti-social behaviour."
3. "An offender below the age of 18 years is delinquent."

This is the legal definition of juvenile delinquency.

According to CBI, Maharashtra among the states and Delhi among Union Territories top the list of Juvenile delinquents or child offenders.

The incidence of delinquency is rising amongst the girls also. You cannot simply say that it is common amongst boys only.

The forms of delinquent behaviour are :

(a) Loitering.

(b) Loafing.

(c) Stealing.

(*d*) Pick-pocketing.

(*e*) Truancy.

(*f*) Gambling.

(*g*) Sexual offences like eve teasing.

Four elements in the social situation seem to be of utmost importance in influencing the behaviour of delinquents—the home, the peergroup, the neighbourhood and the school, political order.

1. The Home and Delinquency. Almost all research workers have accepted that homes of delinquents are characterized by discords, desertions and divorces. Such homes have been pointed out as one of the main causes of delinquency.

More damaging than the absence of parents is the presence of intrafamiliar tension. The poor child has divided loyalties, he knows not to whom he should belong. Consequently he withdraws. All such children are unloved insecure and ill-supervised from peergroups. These peergroups when highly organized turn into gangs. These gangs, later on, indulge in anti-social activities. Many investigators feel that fundamental causes of delinquency are to be found in the treatment accorded to a child by his parents, especially during pre-school years. The basic theme is rejection. When both the parents are alive, a poverty stricken home presents tremendous handicaps to the rearing of children as well-adjusted individuals. The crippling compulsions of economic stringencies contribute to the unfulfilment of basic needs. Small families mean more delinquency.

2. Neighbourhood and Delinquency. The immediate environments of a child also affect the trend he will adopt in connection with his personality. It has been seen that more delinquents come from slums and thickly populated areas. A slum is not merely an aesthetically unpleasant place, it is an area loaded with dynamite for entire community. The newspapers that glorify crime focus the attention of children on crime and furnish them with rationalizations for any later misconduct. From newspapers, the children learn about inconsistent moral values of adults, of business dishonesties, of black-markets and political corruption.

3. The Peergroup and Delinquency. To those in sore need of a substitute for family-love and group belongingness, the peer-group or gang presents itself as a kind of closely kint unit that will answer the purpose. Recognition achieved through being noticed in a particular group is a good substitute for being loved by a parent. The delinquent may dramatize himself as a hero before his person in the dangerous enterprises on which they embark. Here, have will find the anti-social sub-culture patterns with which his feelings agree.

4. School and Delinquency. Although schools are playing an increasingly important role in the training of future citizens, they are also contributing in many cases to juvenile delinquency. The chief adverse element is the nature of the curriculum. Delinquents are typically non-bookish, non-verbal and non-academic individuals who do poorly in traditional school subjects. It is probable that young delinquents experience a great deal of frustration in the course of their school life.

School discipline is sometimes administered as to be thoroughly unacceptable even to well-balanced normal children and adolescents. The effect on delinquents is disastrous and it stimulates them to even greater hostility to school.

Teachers also influence students through emotional atmosphere of their classroom. If they are harsh, domineering and authoritative, they arouse the aggressive hostility of the already rejected child who now finds himself rejected once more. Perhaps the worst teacher for the delinquent in the making is the one who not only cannot accept him but also takes the reaction to his rejection as a personal insult.

4. Democracy and Delinquency. Democractic order increases delinquency. Look at the extent to which things went before the emergency was declared in India. And then we boast of Youth Organizations. Without reflecting on any one, it is said that youth organizations legalise delinquency or delinquent behaviour. Moral pressure can help in curing this. Lawlessness is at its extreme now!

Remedies for Delinquency

It is important to detect the potential delinquents early to treat them before they become problems for the society. Public awareness regarding delinquency had not been fully aroused in India. It is hard to reform a delinquent because his mode of life is extremely satisfying. It provides him what he wants with least possible delay, it brings him prestige among his fellows, it permits him to punish those who have neglected and rejected him. Still various projects of individual and group-therapy have proved quite successful. The child guidance clinics can be established for dealing with delinquents. Juvenile Jail at Bareilly (Uttar Pradesh) is one of the best in India. But prevention is better than cure. Naturally parents and teachers cannot be solely responsible for the prevention of delinqueney. The police, local authorities, the cinema, the newspapers and other literature, all have a part to play. However, parents and teachers can help in many ways.

Following are some of the suggestions in this regard:

1. Affection and Security. Accept a delinquent as a person in his own right. Give the child affection and security.

2. Reject Delinquent Behaviour and not the Delinquent. Reject the delinquent behaviour without rejecting the delinquent. A small amount of genuine affection will often do more to help a boy rehabilitate himself than all treatments combined.

3. Variety of Experience. Provide a child with a variety of experiences, crafts, art, music, athletics, covering a wide range of difficulties and interests can serve the purpose.

4. Early Treatment of Maladjustment. Watch for signs of maladjustment. Early treatment may prevent this maladjustment from taking a delinquent trend.

5. Handle Delinquents Firmly. When an out-burst of delinquent behaviour occurs in class, do not get disturbed, handle it objectively and try to get into the delinquent's world and see things from his point of view.

6. Ignore Minor Incidents. Accept minor instances of bad manners without comment and provide opportunities for normal emotional responses.

7. Improvement of Surroundings. Do what you can to change the conditions of home, school and community that seem to give rise to such behaviour.

8. Stable System of Values. Attempt to build up a stable system of moral and social values.

9. Give Responsibility. Give a potential delinquent some post of special responsibility such as task of preventing other children from committing delinquent acts.

10. Educate the Child. Once a delinquent act has been detected, never pass it over. Make clear to the child that he has done wrong and, if necessary, punish him in an appropriate manner. But once any punishment has been given, accept the child again as the member of the community.

11. Encouragement to Admit Faults. Encourage the child to talk about and admit the existence of anti-social tendencies.

12. Minimize Temptations. Minimize the chances of a child's going wrong by putting the smallest possible number of temptations in his way.

EXCEPTIONAL CHILDREN

Article 45 of the Indian Constitution directs that free and compulsory education should be provided to all children including those who are physically, mentally and socially handicapped. Till now, no steps have been taken to extend the provisions of compulsory education to physically, mentally and socially handicapped children. Therefore, it is our moral responsibility to see that a positive educational programme is devised under which facilities would be available for the education of these specific groups including those who are gifted.

Teaching the Deaf

The number of deaf schools in India is negligible. Loss of hearing may be present at birth or may be acquired at any period during a person's life span. The degree of loss may vary from slight to profound and the effect of the loss may manifest itself in varying degrees in personal, emotional and social adjustment, in educational achievement and vocational efficiency. The earlier in life the loss of hearing is identified the sooner the individual can be given the special medical and educational assistance which may help him to assume his place satisfactorily in the society of which he is an integral part. On the basis of the research evidence the following groups are classified:

1. **Children with Slight Losses.** These children are on the border line between normal hearing and significant defective hearing.

2. **Children with Less Marked Losses.** These children are on the border line between the hard of hearing and the deaf. They do not have enough hearing to learn language and speech with unaided ear, but have residual bearing which can be utilised in their education.

3. **Children with Marked Losses.** These are hard of hearing children.

4. **Children with Profound Losses.** These are the deaf children who do not learn speech and language through their ears even with benefit of amplified sound.

Children who fall under category A usually need no special consideration other than favourable seating in the classroom. Children under category B should be provided with hearing aids and should receive auditory training. Special class instruction may be provided for the more severely hard of hearing in this group, if the adjustments are unsatisfactory in the regular classroom. Children who fall under category C are the children who suffer from marked losses. Their loses range between 54 and 75 decibels in the speech range in the better ear and who have sustained loss from very early childhood. These children may be considered educationally deaf since they require instruction especially designed to foster the acquisition of language and speech even when they make optimum use of hearing with the help of hearing aid. Given adequate educational opportunities, they may acquire language and speech rapidly. Initially these children should receive training in the deaf schools from teachers trained to develop language and speech. Children who suffer from profound losses, *i.e.*, category D suffer from the inability to distinguish highest measurable level of intensity.

Scientific data pertaining to some phases of the problem of deafness in India are amazingly lacking. Statistically speaking, we know comparatively very little about the extent of deafness in India. Therefore, the first step should be to have complete data gathered about the deaf in India. In each Tehsil at least one teacher should be engaged for the service of children who have marked hearing losses. They should have received training in the deaf schools established elsewhere. The states of India should nominate at least 10 persons each year for receiving training in deaf schools whose total number in India is 44. This special training will provide a thorough under standing on the principles of hearing rehabilitation; testing of hearing educational diagnosis; lip reading and auditory training; speech development and correctional; language development, reasoning technique and extensive practical training with deaf children.

Teaching the Orthopaedically Handicapped

Orthopaedically handicapped children include all children with defects in size and structure of bones and joints with deviation in muscle strength, coordination or control. Whenever their physical abilities permit, orthopaedically handicapped children should attend regular classes with normal children, where they would get almost the same kind of experiences as possible and may learn to respond on the level that will be expected of them when they enter the competitive world. For those who are not physically capable of such a programme, it is desirable, as well as necessary to provide facilities, curriculum and equipment so that all may receive educational benefits. Residential facilities should be given to those who suffer from acute handicaps. To those special teachers who know the technique of handling such type of children facilities should be given. Moreover, transport facilities should be provided for whom transportation facilities are not available, or who are excluded from school for some other reason. For all types of orthopaedically handicapped children use should be made of visual aids, motion pictures, slides, stereopticians and projected books which may help the child to read. Interest in music should arrange group projects in art, venturing into story telling and singing. Finger painting should also be explored. Extra-curricular activities which are closely linked with various places of class-work and at the same time provide excellent means of outside social contacts including outings school picnics, trips to zoos, museums, factories, farms, etc., should be arranged.

Vocational guidance, education and training are integral functions of training programme which the authorities should take note of. Also psychological service should form an effective part of the educational programme for the orthopaedically handicapped children. The most important function of the teacher is to help the child to accept his handicap, to prevent the psychological crippling which is more damaging than the physical handicap. He should possess personal friendliness, warmth, interests, patience, kindliness and honest liking and respect for children. Every effect should be directed towards assisting the child towards the greatest possible degree of self-acceptance, self-reliance and adjustment to his limitations. Teacher must also understand the family background of the handicapped child.

Teaching Children with Speech Handicaps

Speech defectives comprise the largest single group of handicapped school children. Speech is the most used of all the language functions and the most fateful, day in and day out, in the social relationships of people everywhere. What is done about speech and especially speech disorders, in our schools is, therefore, of the utmost importance to the pupils as individuals and to the society which they will help to develop as they become adult citizens.

Stuttering affects six to ten out of every thousand school children. Stammering children comprise probably the largest single group of handicaps in our schools. But one thing is certain that for the most part they are normally educable and potentially employable aside from their speech defects. In view of the substantial progress made in speech correction in recent years, the great majority of handicapped children can be corrected or materially benefited. Again, much of the special attention and help handicapped children need can be given by the classroom teacher through judicious adjustment of teaching policies and methods, together with special instruction and personal attention that is not exclusively time consuming and can be given by any teacher who has had an elementary education in speech correction.

Mentally Handicapped Children

The problem of a child with retarded mental development assumes considerable importance in the general programme of education. Authorities as well as teachers have to become increasingly concerned about those pupils who find it difficult to profit from the ordinary programme provided for normal children. In the absence of special provisions for the mentally handicapped, the problem becomes acute because the curriculum is not adapted to the needs and abilities of these pupils. No survey on a large scale has been conducted in this country, much less in this State, to find the incidence of mental deficiency. The surveys conducted so far indicate that six lakhs of children population in the country may be mentally deficient. Gokhale gives the estimate of mentally handicapped children in India as 15 to 18 lakhs. This is the most serious problem which requires serious thought and serious action by all sections of society. Some of specific characteristics which may be noted in determining mentally handicapped children are: Social incompetence resulting from developmental mental arrest of constitutional (heredity or acquired) origin; the condition is essentially sometime incurable through treatment and unremedial through training though in favourable circumstances mentally handicapped may improve.

Programme for Socially Maladjusted

In developing programmes for socially maladjusted children, research has shown that certain principles should be followed. Some of these have been suggested in the preceding pages. They are restated here for the sake of emphasis. The following statements will be found vague by those interested in the field of educating and caring for maladjusted children. They should be considered in connection with the task of organising a special education programme in this field.

1. Socially maladjusted children are entitled to the advantage of a special educational programme that will permit them to develop to the limit of their capacities. All children must have the right to develop into self-respecting, useful citizens by the process of public education and that right must not be abridged by a handicap of any kind which can be eliminated or mitigated through the facilities and resources of the schools.

2. In organising and administering a programme of special education for the socially maladjusted school, administrators must maintain a balance between the interests of pupils needing placement in special groups and the interests

of the great majority of the school population. When these interests often conflict, the conflict must be resolved for the best interests of an concerned. In general, placement of any child in a special group should not be made if that child may receive as good or better training in a normal group, even though it may be necessary to give special help and additional services over and above those which are usually provided.

3. Segregation as commonly defined is not a necessary concomitant of the education of socially maladjusted children. School administrators should realize that an exceptional child may be more harmfully segregated when kept in a regular class which meets his needs than when assigned to a special class which meets his needs much better.

4. The education of socially maladjusted children requires a broader basis than that of mere intellectual development. These children often have warped personalities and consequently, their feelings and attitudes are the object of more concern than their academic attainments. Educational programmes for meeting the needs of the socially maladjusted should be subject to a minimum of regimentation and should have more than yearly opportunities for change. Children who are deviates because of social maladjustment need a chance to develop emotional stability; they need personal, educational and vocational guidance; they need to experience the sense of security that goes with a socially acceptable personality.

5. Any programme of education for the socially maladjusted will be considered by the selection of properly qualified and trained personnel. The teacher must have personal qualifications suitable for the task in hand and should have training which is particularly adapted to the requirements of working with problem children.

6. No programme for socially maladjusted children is sound unless it recognises the fact that the behaviour of such children is symptomatic and purposive. Adverse behaviour

such as truancy, incorrigibility and delinquencies are only symptoms of underlying conditions, the roots of which will be found in the environmental life of the child or in some physiological and psychological aspects of his personality. Teachers and school administrators must also recognise that so-called abnormal behaviour is sometimes to be regarded as logically normal. An objective attitude on the part of teachers toward children's behaviour may serve to prevent problem cases from developing.

7. School systems should provide for early identification and early diagnosis of children who are maladjusted. School children become problem cases as a result of experience and at a much younger age than is often thought to be the case. While truants, incorrigibles and delinquents become acute problems at about the time of adolescence, the years in school previous to that period should be considered years of opportunity.

8. Socially maladjusted children differ from normal children more in degree than in kind. There is no hard and fast line between normal and abnormal adjustment. Problem boys and girls look no different than ordinary children. Too often problem children are not problems in their relation to society. They are, in fact, only children with problems which they their parents and their teachers cannot readily solve.

The degree of concern on the problem of gifted children is growing. There are legitimate reasons for it. Guiding and growing the potentialities in children is the primary concern of education and no country, however big or small, can ignore this problem.

Due to some unfortunate reasons, there have been some prejudice regarding gifted children in some quarters in the past. They have generally been looked down upon as bespectacled old and angular people who have a strange set of behaviour patterns and who are difficult to understand. All types of names and or titles have been attributed to them.

Such assumptions about them have often been misleading and distorted. It has often been said that gifted people are queer and maladjusted and have neurotic and schizophrenic tendencies in them. It has, however, been difficult to prove such assertions.

First studies which have drawn importance to the problem of gifted children are those of Terman. His studies have provided objective information about the psychology and physiology of gifted children. These studies have removed many misunderstandings regarding gifted and creative people. Now it is known, without any prejudice, that a gifted child, is generally more healthier, more attractive, handsome, better developed and in general, to be enjoying a richer and fuller life. After his studies, more and more information is available about gifted children. Their talents, their attributes, their characteristics are better appreciated now. Within the past three decades, selection and education of gifted children has become a major job function of any normal system of education.value of the problem at hand.

Identification of Gifted Children

Identification of gifted children is a problem that has engaged the attention of psychologists and educationists all over the world. Giftedness is a psychological and educational problem. There are varieties of giftedness in children. It becomes, therefore, difficult to assess giftedness with the help of a single tool or test.

Terman's model for identifying gifted children was both simple and powerful. He employed group or individual tests for assessing giftedness in children. Children in the top 1 or 2 or 3 per cent in I.Q. were called by him as gifted children. It was, however, felt that common intelligence tests represent a rather narrow band of intellectual tasks. Such tests usually assess recall or recognition type of problems. Hence, full range of giftedness in children cannot be measured by intelligence tests fully. Moreover, it is also believed that children who score high on intelligence tests may not necessarily be highly creative also. It was also felt that IQ units may not be necessarily immune to advances in one's understanding of thinking and behaviour. This fact or conclusion is much vouchsafed by result from studies done in the field of cognition, learning and problem solving. There are studies which have suggested that creativity and intelligence might not be closely related. Further, intelligence tests emphasise learning ability and school achievement and may ignore or neglect those psychological functions which are related to creativity. This way of thinking has been emphasised by Thurstone who has stated as follows:

> "To be extremely intelligent is not the same as to be gifted in creative work. This may be taken as a hypothesis. It is a common observation in the universities that those students who have high intelligence, judged by available criteria, are not necessarily the only ones who produce the most original ideas. All of us probably know a few men who are creative and highly intelligent, but this combination is not the rule."

Observation, given above, will lead scientists in their attempt to identify gifted children in the school population. Once it is clear that intelligence and creativity may not be found in a child in the same measure, the level of two abilities have to be fully ascertained.

Intelligence tests and creativity tests have been used for identifying giftedness in children. In general, tests of creativity involve the ability to deal inventively with verbal and numerical symbol systems and with object-space relations. Performance on these tests depended not on a single predertnmined correct response as is most often the case with general intelligence tests, but on the number, novelty and variety of adaptive responses to a given task. Getzels has described the following measures of creativity or giftedness:

1. Word Association Test. The test presents the subject with words and each word and as multiple meanings. The student is asked to write as many meanings as he can for each word.

2. Hidden Shapes. The subject has to identify the complex figures in which the single figures appear.

3. Uses for Things. The subject has to write as many different uses as he can for each object.

4. Fables. The subject is given fables whose last lines are missing. The student is required to supply a suitable ending.

Scholastic achievement tests have been employed by workers for studying giftedness. But such tests have not been found comprehensive or valid enough to assess creativity.

Efforts have been made to study giftedness from another angle. Kough and De Haon have developed a method for discovering special abilities and disabilities. The authors have listed various criteria to identify special abilities and talents among gifted children. The criteria fall in the following areas of human endeavour:

1. Intellectual ability.
2. Mechanical skills.
3. Physical skills.

Another observation technique called "Guess who technique" has also been employed by them to identify intellectually gifted children in the student population. Another author, namely, Willy has used the following procedure for identifying giftedness:

1. Accuracy and use of vocabulary.
2. Quick and keen observation and retention of information about things.
3. The early development of ability to read.
4. Language proficiency.

5. Quality of concentration.
6. Early interest in calendars, in telling time and in clocks.

Intellectual Characteristics of Gifted Children

Gifted children are endowed with many personality and intellectual qualities. Teachers and parents have provided significant information about mental and personality potentialities of gifted children they are:

1. They display rich common sense and practical knowledge.
2. They are ahead of their class fellows.
3. They are question-oriented.
4. They are logical, think coherently and recognise relationship.
5. They have good memory.
6. They have a good vocabulary.
7. They have good fund of general knowledge.
8. They are alert, keen, observant and react quickly.
9. They use good and unusual ideas.
10. They learn rapidly and easily;

Case studies of the intellectually gifted children have above characteristics in common. Life histories of gifted people have shown to possess many outstanding qualities. Gifted children come generally from homes which have good socio-economic level.

The physical characteristics of gifted children are conspicuous. It has been found that in physical and general health, gifted children possess better standards generally than an average child. They are heavier in weight and their height and vitality is also superior. Their reactions are quick and they develop fast. They enter school at an early age and in school they are much ahead of their class. Their activities in classrooms are wide and varied. Their interests are also very diverse. The gifted children are found to be aware of their talent and their utilization. Their cognitive processes are rich, their memory superior and their divergent thinking high. They show the process of sensing gaps in problems and are quick in filling missing elements intellectually. They form ideas and are getting away from the main track. They try to break away from the mould and are open to experience. This initiative in them to break away from the usual square of thought into altogether different pattern of thought gives facility to invent. They search, combine and synthesize in their minds. Other attributes of their mental activities are marked by the following characteristics:

1. They possess fluency and have most ideas in them.
2. They possess originality and unusual ideas.
3. They show flexibility in their mental thought process. They are first to find a new way.
4. Their mind is inventive and they invent and develop ideas.
5. Their mind has the power of elaboration by virtue of which they think of all consequences.

They have ability to retain ideas and to call for innovative and constructive thinking. A gifted child may enter the school at usual age, but he learns quickly and is ahead of the class and is liable for academic advancement during his school years. Terman, a psychologist, states that a gifted child is ahead of the class by all means and masters the school curriculum to a point two full grades beyond the one in which he is enrolled and some are as much as three or four grades ahead of other boys.

Characteristics of Gifted Children

Gifted children have been found to have a rich personality. It has been found that there is positive and intimate relationship between giftedness and personality. It is found that gifted children are more desired, better known or understood, more ambitious and more hard-working. They are usually having strong urge to explore and to create. They are also found to be able to sustain frustration better. They self-initiate learning. Their temperament and motivation are better as compared to average children.

It has also been found by various authors that gifted children are impulsive, more self-confident, appreciate originality in ideas. They are highly interested in aesthetic expression and in meditative or reflective thinking. Other characteristics of the personality of gifted children are described in the following way:

1. They accept order.
2. They like solitude and are individualistic.
3. They have desire to excel.
4. They are adventurous.
5. They are attracted to mysteries.
6. They are fault-finders and full of curiosity.
7. They shun power, are sincere, visionary and versatile.
8. They are determined, discontended, emotional and energetic.
9. They are intuitive, industrious, reserved and self-starters.

Other significant personality characteristics of gifted children as found from various studies are:

1. They have strength of self-image.
2. They are humorous.
3. They have greater-awareness in them.
4. They can recall easily.
5. They have creative acceptance of oneself.
6. They have uneven edged development.

Gifted children are possessed by a purpose which is worthy of enthusiastic devotion. They have high degree of motivation. They have been found to have a sense of psychological isolation and estrangement from others. They have developing values and purpose.

Gifted children are fully capable to use their unconscious. Psychologists have stressed the role of unconscious in giftedness. Gifted children are able to carry pre-cocious thinking successfully. They have high motivation for achievement and are highly sensitive, resourceful flexible and willing to get off the beaten-track.

As said above, conflict plays a great part in creativity of gifted children. The bases of conflict is in unconscious. It is said that creative solution and neurotic solution to problems have the same genesis. Unconscious forces, motivating the gifted solution are parallel to the unconscious forces motivating the neurotic solution. As the instructual pressures rise and a neurotic solution appears imminent, the unconscious defence against it leads to the creation of an art-product.

Role of childhood experience in gifted or creative product is emphasized. Gifted or creative behaviour is seen as a continuation and substitute for the play of childhood. The creative thoughts are derived, it is said, from the elaboration of the freely rising fantasies and ideas related to day-dreaming and childhood play. The psychic function and effect of creative behaviour is the discharge of pent-up emotions resulting from conflict until a tolerable level is reached.

Education of Gifted Children

Education for gifted children is one of the important problems to which educators and administrators have to administer themselves. Development and abilities of gifted children has to be attempted through an educational system. For this reason, teachers, counsellors, administrators and others should be familiar with requirements of education for such children.

They have also to know the developmental aspects of giftedness so that educational process is geared to the teaching of superior children. Education for gifted children, however, is not an easy task as it involves many agencies and organizations. In a national system of education, provision for the education of gifted children has to be made carefully so that gifted children may not feel frustrated in classrooms on account of want of proper stimulation. Children in a particular class may not have the same level of intelligence. This warrants introduction of special classes for those children who are gifted with high abilities and aptitudes.

Regarding acceleration, the author feels that gifted children should be admitted to classes on the basis of mental age and not chronological age. This may also involve of keeping of grades for them. Since skipping of a particular grade may leave a gap in a child's experience, child may be exposed to various educational experiences in a good manner. This method is called telescoping of grades. Extra coaching and or extra classes may attend to this sort of acceleration in learning.

Enrichment is also accepted as a useful alternative to educational system for gifted children. Gallagher defines enrichment method as "the type of activity devoted to the further development of particular intellectual skills and talents of the gifted children. According to Kirk, this enrichment programme in education for gifted children can be run in the following way:

1. Gifted children should be engaged in additional readings and extra assignments.
2. Gifted children should be expected to give high results in school performance.
3. Gifted children should be combined in a group so that they stimulate each other in learning.
4. Gifted children should be given a special programme of curriculum to follow in their schools.

5. Gifted children should be allowed to participate in large-curricular activities.
6. Gifted children should be allowed to participate in activities that encourage initiative, independence and creativity.

Role of early schooling for gifted children is very important. It not only gives early discipline but early training to them to think affectively. This education should sustain curiosity and manipulativeness in them. A natural system of education for gifted children should prepare them for victory and success and not for frustration and failure. Gifted children should be encouraged to think creatively so that they work about machines and invent imperative human relations, the capacity for weighing of values and develop consciousness in them to liberate the society from stagnation.

The education for gifted children at primary and secondary levels have to be different from average children. While acceleration, enrichment and special education may not be necessary for an average pupil, for gifted children it is very much necessary. A special programme of education, especially at the elementary school level, has to be provided for gifted children, who have to be more content either in a homogeneous or in a heterogeneous class or be allowed to have opportunity for faster coverage.

Since content of schooling is a matter of great importance for the school run for gifted children, the curriculum becomes one of the important areas where requirements of a gifted child should be reflected adequately. Moreover, the curriculum must also reflect the potentiality of satisfying the individual abilities and aptitudes of gifted children. Accordingly, high school curriculum must be wide and comprehensive and be structured in such a way that it provides a wide and reasonable variety of educational experiences to the gifted child.

Kirk refers many adaptations in educational activities of gifted children so as to sustain and stimulate them in the school. Some of the adaptations recommended by him are:

1. Increase the counselling and guidance activities in school for gifted children.
2. Extra courses should be permitted for the gifted children.
3. Offer extra-curricular activities—school particulars, science clubs, hobby clubs, student government, etc., to gifted children.
4. Advance classes in science, mathematics, English, social studies should be organized for the gifted children. These classes should emphasize on ideas, concepts and relationships and not on mere memory alone.
5. Gifted students should be allowed to enrol in nearby colleges and universities for advanced courses in a subject in which they are interested.
6. Courses for gifted children should be revised from time to time.
7. Special correspondence courses should be initiated for gifted children.

It is evident that the environment in which a gifted child will live should be objective and warm and it should enable him to think and to find his anchors in reality. It should be stimulating but not soft and unrealistic. It should be responsive to the needs and requirements of the gifted students.

BACKWARD CHILD

Generally speaking, no mother or father is easily prepared to accept that their child is mentally backward. But modern intelligence test techniques have indubitably established that some children are mentally backward. A backward child is more in need of assistance than one with average or above average intelligence. Hence, in modern schools teachers are expected to pay particular attention to the weak students in their classes. A backward child, evidently, is one who has an intelligence quotient lower then the average child. In the words of Cyril Burt, "A backward child is one, who in the middle of the school career, is unable to do the work of the class below that which is normal for his age." In this way, the backward child falls far behind other children of his age in matters of study. In school, such children remain in the same class for many years and fail to progress upward. Education psychologists have no single opinion on the intelligence quotient below which a child can definitely be called backward. Yet, generally psychologists classify those children as backward whose intelligence quotient falls below 85.

Children who are backward have certain problems that are peculiarly their own. Having a very low intelligence quoting, they require to be taught many things by others. A backward child cannot lead, he is more in a position to end the help and sympathy of friends and relatives. A child who is very feebleminded, having an intelligence quotient of 70 or less, cannot obtain education in the ordinary school and has to be sent to special institutions. Emotional adjustment of the child is completely disintegrated if he fails in the same class time after time. In a class, teaching a approximates to the average level of students. Hence, it is only natural for the backward child to meet with failure.

Backward child also faces difficulties in social adjustment. In the group he does not easily grasp even the smallest thing and rules of play have to be respected for his benefit. His colleagues naturally heckle him. It is possible, that a spark of leadership may be ignited in him if he communicates with those younger than himself but on the same intelligence level. But there he is unwelcomed since other children steer clear of him because of his physical superiority. Adults pay him little or no attention, instead he is scolded from time to time. In adolescence and adulthood, too, the situation of this child does not improve. Activities of the group are usually on a higher level, so that he does not fit in them either.

Causes of Backwardness

It is not even possible to recount all the various causes of backwardness, but the main causes are the following :

1. Physical Causes. Due to many physical defects and abnormalities, the child lags behind other average children. Such physical conditions would be weakness of the eyes, deafness or being hard of hearing, stuttering, being weak or crippled, etc. This also has an adverse effect upon his social adjustment.

2. Educational Immaturity. Physical and mental short commands in the child result in educational immaturity. Putting it differently, he has no adequate maturity to grasp the teachings is customary for his age leave.

3. Mental Defects. It has been indicated earlire that the backward child has a very low intelligence quotient as compared with that of an average child. Having a low intelligence, his other mental characteristics and abilities also do not compare favourably with the average child.

4. Lack of Interest. Sometimes, even though the child is free from all the defects and shortcomings mentioned above and even those that have not been mentioned, he falls behind the rest of the class for the simple reason that he cannot stir any interest in the various subjects being taught. This causes repeated failure.

5. Emotional Factors. The child's social adjustment is defeated by his mental and physical abnormalities and he meets with successive failures. He meets with criticism on every account. Hence, it is natural for him to manifest anxiety tension, depression and there emotional factors. Sometimes though every other defect is absent, the child is backward in his class for no better reason than these psychological or emotional factors.

6. Atmosphere in the Home. One reason of backwardness can also be the atmosphere in the home. Proper development of the child cannot occur in a virulent family atmosphere. Hence, he falls behind children of his age group. And among the factors that contribute to the defective atmosphere in the home are constant quarrelling and fighting between parents and children, unbalanced diet, strict discipline, absence of all peace and calm, proverty, etc.

Types of Backwardness

Backwardness may be of two types—general and specific backwardness. A situation when a child is weak in almost all the subjects in general is called *general backwardness.* But it sometimes happens that the child is good or at an average level in many subjects except one or two. This is *specific backwardness*. Specific backwardness in a particular subject is often due to weakness in a specific ability needed for that subject. *e.g.,* verbal or numerical ability.

The school authorities should face two facts :

1. That children who are backward owing to low general intellect need to be grouped into speical class and not only taught at a slower pace than the average children but also given more concrete and practical curriculum.
2. Children who are specifically backward in a subject say arithmatic, reading etc., need individual teaching in that subject and often it is necessary to start again from the very beginning.

The possible causes for turning a child into backward may be divided into two broad categories - internal, lying within the individual and / or external, lying in his environment.

(1) Causes Lying within the Individual Child. The child may become backward on account of the factors or things lying within his body structure as:

(*a*) Child may have poor health or suffer from physical ailments and chronic diseases etc. and it may obstruct him to harness the actual potential of his natural abilities.

(*b*) Child may suffer from some bodily defects like defective vision, faulty hearing, speech defects, left handedness etc. These defects or improper functioning of the body organs, particularly connected with sense organs, may prove a stumbling block in attaining what could have been otherwise expected from him.

(*c*) Child may not be utilising his intellectual powerss like power of concentration, reasoning, thinking, observation and imagination etc., as he should otherwise do for actualising his basic potential on account of some defects in his brain system or its functioning or on account of some emotional and psychological pressure generated as a result of his personal and social maladjustment.

(2) Causes Lying in the Environment. The external environment may also force the child to turn into backwardness like:

(*a*) ***Defective Home Environment.*** The uncongenial and improper environment prevailing in a family or home has a negative force in the educational progress of a child. Such negative factors present at one's home may be termed as:

(*i*) Tensions, conflicts and quarrels at home between the parents and among the members of the family.

(*ii*) Poverty, illiteracy and intellectual inferiorty of the parents or gaurdians.

(*iii*) Improper behaviour with the child due to over protection, negligence, jealousy, rivalry etc.

(iv) Over engagement of children in household activities or family occupation.

(v) Noisy environment and inadequate physical facilities for the study.

(b) *Defective School Environment.* Conditions prevailing in the school with regard to men and material facilities and the child's adjustment to the available school environment contribute much in the attainment of what is allowed to the child by his natural abilities. The adverse and unfavourable conditions and improper environment lead the child to educational backwardness. These unfavorable conditions and factors may be of the following nature:

(i) Lack of discipline and poor administration.

(ii) Lack of proper guidance and counselling services.

(iii) Lack of trained and qualified teachers.

(iv) Defective curriculum, uninteresting and wrong methods of teaching and lack of proper evoluation system.

(v) Lack of material facilities like as of accomodation, furniture, library, laboratary, teaching learning aids and equipments and other things helpful for co-curricular activities.

(vi) Lack of proper attention for the regular attendance and supervision of the drill and homework.

(vii) Indifference or improper behaviour of the teacher.

(c) *Other Negative Social Influences.* Apart from the defective home and school environment there may be other negative influences and undesirable forces present in the surrounding school environment of the child. He may have an adverse and unfavourable neighbourhood, evil company of the playmates and other members of the society and day-to-day quarrels and disturbances perturbing his peace of mind. He may be tempted by what is available in his social environment and may pick up bad habits like smoking, playing cards, gambling, taking drugs, reading improper and cheap literature, seing too much movies, playing truancy at the school hours and watching excessive T.V. etc. All these things, perpetuated by the negative environment influences, are quite forceful to affect the whole life style of an individual child including ways of thinking, motivational process and habits of work and studies.

✱✱✱

10

Guidance and Counselling

Guidance process is a purposeful process. Without determining the objectives, this process cannot be completed or concluded. Due to the variations in the scope, guidance has a variety of objectives. If we start guidance process without any objective it cannot be concluded because various sub-processes involved in the guidance cannot be given any direction and these activities ramain meaningless.

Objectives of guidance change at each level of education. Every stage of education enjoys different objective of guidance. *For example,* the functions and objectives of guidance at primary level will be different from those of middle stage. Similary, objectives of guidance at high school or higher secondary stage will be different from those of middle stage:

Objectives of Guidance at Primary Stage

Every activity at primary stage includes guidance. At primary stage, there is need to pay special attention towards the health of the children. At this stage there is great need to co-ordinate the functions of a teacher, guidance worker, school, staff members of medical service and school social service workers. All these persons help in learning properly essential skills for primary education, for acquiring basic knowledge, for adapting according to the social traditions and for developing proper attitude towards school activities. These too are the objectives of guidance at this stage.

When a child leaves home and enters into the school environment, he has to make many co-ordinations. School environment is very comprehensive for a child. In school, the child comes in contact with various types of persons, children and teachers. Also, he lacks security in the school and he becomes very fearful, hesitating and remains suppressed in the school. Those children are more in difficulty who are fully dependent on their parents.

Primary stage ranges upto fifth class. During this period, children make their place in the school. Gradually they get themselves established in the school and they consider themselves an honourable member of the school. When these children cross over from primary stage to higher stage, they feel suffocated. Behaviours of primary stage sometimes look like a joke. The traditions of new school, rules, administration, curriculum etc., are all new. In such situations, the guidance worker plans the guidance programme in order to prepare the children of primary stage, for next higher school stage, such as arranging the visits of primary stage children to high schools, inviting high school teachers to primary schools, explaining the procedure for getting admission to high schools after leaving the primary schools, making them familiar with the curriculum of high schools etc.

In nut-shell we can say that the following are the objectives of guidance at the primary school stage:

(a) To co-ordinate all the activities of all the workers such as a teacher, medical service staff members, school social-workers and the activities of a person providing guidance etc.

(b) To assist developing higher attitude towards school activities.

(c) To assist children in developing their physical and emotional stability.

(d) Objective of finding out problems regarding adjustment in the school and to control them.

(*e*) Objective of creating the feeling of co-operation in the children.

(*f*) To assist children adapting according to the school traditions and rules-regulations of the school.

(*g*) Objectives of making children ready for crossing over from primary school to next higher stage.

(*h*) Objective of supplying various informations to children regarding higher stage such as curriculum of next higher classes and distribution of booklets regarding various school functions among the children and their parents.

Objectives of Guidance at Secondary Stage

The child leaves primary stage and enters secondary stage. At this stage the scope of guidance becomes more comprehensive as compared to the scope of guidance at primary stage. At this stage, guidance becomes more important as a result of multiple reasons. At this stage, teachers too teach according to the subjects. Therefore, all the children cannot contact all the teachers. At high and higher secondary schools, only subject specialists teach. At these stages, personal, social and vocational problems of the children come into light and solution of these problems is necessary. In order to achieve this objective, a comprehensive and organized guidance service is continuously required.

In this way, we see the following are the objectives of guidance at secondary and higher secondary stages:

1. Objectives of Helping in Selecting the Subject. After reaching at this higher level, the selection of subject becomes necessary. In this task, the pupil may face difficulty. He is unable to understand in which subject he is competent and which subject he should not opt. The selection of subjects has become very difficult task due to the variety of subjects. Hence, the objective of guidance should be to understand this type of problem and the same should be solved.

2. Objective of Fulfilling the Needs, related to Students Health. The objectives of guidance service existing in the school is to take care of needs related to the health of the children and to inform timely the relevant persons regarding those health deficiencies so that the necessary action can be taken.

3. Objective of Familiarizing with New School Life. A child is not familiar with the life of new school when he leaves the primary school and enters to a new higher school. He has to face a number of problems regarding admission. Also, it may become difficult for him to co-ordinate with new classmates and new friends. Hence, the objective of guidance service existing in the school should be to eliminate such problems of the pupils.

4. Objective of Co-curricular Activities. Co-curricular activities are being considered very important for the physical development of the children, such as debates, sports, cultural activities, programmes of social service etc. Hence, the important objective of guidance services is to guide the children for such activities so that their physical development may be given proper direction.

5. Objective of Creating Environment. According to Crow and Crow, the creation of academic environment in accordance with the adolescent stage is very essential so that the pupils may be motivated in order to acquire higher education and they may be helped in selecting the subject according to abilities.

6. Objective of Maintaining the Records. In schools it is important to maintain the records regarding the development of the pupils, education, psychological development, cultural activities, because the entire guidance programme is to be conducted on the basis of these records. Hence, the objective of guidance programme workers should be to maintain such record so that the counsellor may feel convenience during counselling.

7. Objective of Co-operation. In solving the problems of the children, co-operation plays an important role. This co-operation is desirable in guidance at all levels. The objective of guidance services should be to create a feeling of co-operation among all the workers involved in this service. This co-operation acts as a substantial base for guidance at the next higher-level.

8. Objective of Emotional and Individual Needs. When an individual enters adolescence, an individual experiences emotional and personal needs. Therefore, the objective of guidance services should be to solve the problems related to such needs.

Objectives at College and University Stage

When a child enters the college he becomes a complete young man and his personality acquires has own pattern. Some of them have clear objectives. Some start feeling their responsibilities. They are also very serious in their studies because they know the meaning of success and failure. But there are some persons who fail to understand what the college education means. They come to college without knowing what they can do. After securing admission to the college they face multiple difficulties. They consider their studies meaningless. That is why they go to college without any objective.

In such situations, the objective of college guidance programme is that the needs of such students should be looked into immediately who are unable to make progress in the college affairs and who can use their telent in their activities on providing proper educational facilities.

In college, the objective of guidance service should also be to help the pupils in selecting the subject so that they can acquire specialization because such specializations help the pupils in their future vocational plans.

Similarly other objectives of guidance service include organized programmes of tutorials and providing help to the pupils for making progress by studying the books existing in the library. It would be very meaningful if the students are made aware of objectives, nature and scope of college educations before its commencement in the college. According to other objectives of guidance service, the pupils should be informed regarding the various facilities existing in the college such as library facilities. If such is not informed, it would lead towards the wastage of pupil's time.

In this way, we see that there is no much difference between the functions and guidance objectives of high, higher secondary and college stage. Hence, the guidance objectives at college and university stage can also be presented in the following manner:

(*a*) To make available informations regarding admissions to college and university.

(*b*) To arrange for testing the pupils.

(*c*) To provide informations regarding co-curricular activities in the college and university.

(*d*) The help the children eliminating their financial difficulties.

(*e*) To make aware the pupils regarding the existing hostel facilities so that the problems of their boarding and lodging can be removed.

(*f*) To help pupils selecting subjects so that they may achieve their future programme and objectives.

(*g*) To provide vocational informations to the pupils so that they may take decision regarding their future.

The objectives of guidance can also be expressed as follows:

(*a*) Objectives from pupils' point of view.

(*b*) From institution's point of view.

COUNSELLING

The scholars have exhibited variations in their view-points regarding defining the term "counselling". One major cause of these variations is—occurrence of historical changes in the perceptions of the persons. Inspite of this, there was a lot of variation in the philosophy of experts.

According to Wrenn (1951), "Counselling is a dynamic and purposeful relationship between the people in which procedures vary with the nature of the students' need, but in which there is always mutual participation by the counsellor and the student with the focus upon self-classification and self-determination by the students."

Ruth Strang, emphasizing mutual relationshps of persons, has said, "Counselling is a face-to-face relationship in which growth takes place in the counsellor as well as the counsellee."

According to Shostorm and Brammer (1952), "Counselling is a purposeful reciprocal relationship between two people in which one a trained person, helps the other to change himself or his environment."

Wolberg (1954), has considered counselling as a form of interview. He has defined it as, "Counselling is a form a interviewing in which the client is helped to understand himself more completely in order to correct an environment or adjustment difficulty."

Myers has written as, "Counselling means a relationship between two persons in which one person provides special assistance to the other."

Rogers says about counselling that, "Counselling does not mean providing a service by means of which specific problem of an individual may be solved. The process should help the counsellee in gaining claer insight into the problem and in achieving the self-confidence."

Considering counselling as a problem-solving effort, Hamrin and Erickson have said that, "Counselling is an attempt made by the counsellee to think through and solve the problem that confronts him. It is an experiment in co-operative thinking, who two individuals (the counsellor and the counsellee) srtike to solve a problem."

Considering counselling as a mutual learning process, Willy and Andrew have said, "Counselling is a mutual learning process involving two individuals—one seeking help and the other a professionally trained person helping the first to orient and direct himself towards a goal which leads to his maximum development and growth in his environment."

Characteristics of Counselling

The following facts become distinct about counselling after studying and analysing the above mentioned definitions:

1. Counselling is a conversation with some one regarding some problem.
2. Normally, not always, one of the two persons possesses facts or experiences or abilities which the other lacks.
3. Counselling helps the pupil in making selections and following those selections.
4. Counselling is a learning oriented process.
5. Counselling is a professional service.
6. In the process of counselling, the problem is made clear through discussion. The counsellor explores the problem and its importance through his skilful questioning.
7. Counselling involves interactions in which the counsellor accepts the responsibility of positive contribution in the development of other person's personality.
8. Counsellor draws out the facts from the pupils through counselling process.
9. He also informs the pupils certain facts.

10. Counselling is only one aspect of guidance.
11. Counselling is an assistance to the persons in their behaviour related problems in which their emotions and motivations are main.
12. Counselling is a face to face relationship with a person. This relationship is between a counsellor and a client.
13. Counsellor guides the pupils to establish contacts with various factors in the counselling process. He explains the importance of those facts which the pupils consider unimportant.
14. Counselling is a vital part of the entire guidance programme.
15. Best counselling is in the form of the decision made by the counsellee.
16. Counselling is possible in humorous and co-operative environment only.
17. Counselling is based on the appropriateness of counsellor's prediction.
18. Counselling is democratic. It lays down the democratic system. The client can behave as he wishes.

Basic Principles of Counselling Process

According to Mc Daniel and Shaftal, the counselling process is based on some basic principles:

(1) Principle of Acceptance. According to this principle, each client must the accepted as an individual and dealt with as such. The counsellor should give, due regard to the rights of the client.

(2) Principle of Respect for the Individual. All the schools of thoughts of counselling advocate for the respect of the individual *i.e.*, respecting an individual's feelings must be an integral part of counselling process.

(3) Principle of Permissiveness. Counselling is such a relationship which develops optimism and the environment shapes according to the person. All the thoughts accept the relative relationship of counselling.

(4) Principle of Learning. All the assumptions of counselling accept the presence of learning-elements in the counselling process.

(5) Principle of Consistency with Ideals of Democracy. All the principles are associated with ideals of democracy. The ideals of democracy desire to accept a person and want to respect the rights of others. The process of counselling is based upon the ideals of a person's respect. It is a process with accepts individual differences.

(6) Principle of Thinking with the Individual. Counselling emphasizes thinking with the individual. It is essential to differentiate think for whom? And 'why to think'? It is the role of the counsellor the think about all the forces around the client to join client's thought—process and to work collectively with the client regarding his problem.

Directive or Counsellor-Centred or Prescriptive Counselling

Counseller-centred counselling revolves around the counsellor. He tries to establish good relations through the friendship and assistance. In this, the counsellor is very active and he expresses his ideas and attitudes independently. He evaluates the expressions of the client. According to this view-point the counselling leads the interview. In this, the counsellor asks a series of standardized questions. Each question carries a short answer. The counsellor does not allow the development of client's expression and feelings. He leads as an expert, evaluates and gives suggestion or advice.

The chief advocate of this school is E.G. Williamson of the University of Minnesota. Hence, according to this school of thought, the counsellor owns entire responsibility of solving client's problem. In this procedure, the counsellor explores the problem and defines it, diagnoses it and tell about the remedy of the problem.

Basic Assumptions of Directive Counselling

According to Andrews and Willy, the following can be the basic assumptions of directive counselling:

1. Counselling as an Intellectual Process. Any client's intellectual capability is not destroyed as a result of his maladjustment. Hence, counselling is primarily an intellectual process. It stresses upon the intellectual aspects of a person instead of his emotional aspects of the personality.

2. Counselling Objectives as Problem Solving Situation. The objectives of counselling are achieved through problem solving situation.

3. Client's Incapability of Solving the Process. The counsellor assumes that the client does not possess the capability for solving the problem always.

4. Competency in Giving Advice. The counsellor possesses the best training experience and informations. He is more competent to provide an advice to solve the problem.

Hence, we observe that the direct and explanatory methods are suggested. In such a counselling focus should be on the problem instead of the person and the client should cooperate in the entire process. The client is to act under the counsellor, not along with him.

Steps in Directive Counselling

Williamson has described the following six steps of directive counselling:

1. Analysis. In this steps, data and informations are collected regarding the situation or the client which can be used in the counselling process as a valid and reliable base. For this counselling step, the following tools are used:

(*a*) Cumulative records.

(*b*) Interview.

(*c*) Time distribution form.

(*d*) Autobiography.

(*e*) Anecdotal record.

(*f*) Psychological tests.

For the integration of entire data, case-history method is used. It includes family-history, recreational interests and habits etc.

2. Synthesis. Williamson has defined synthesis as—"The process of summarizing and organizing of the data from the analysis in such a manner as to reveal the student's assets, liabilities, adjustments and mal-adjustments."

3. Diagnosis. Diagnosis consists of the inter-pretations of the data in terms of the problems indicated and assets and liabilities, the strengths and weaknesses of the students.

Diagnosis has the following main three stepes:

(*a*) Identification of Problem.

(*b*) Discovering the cuase.

(*c*) Prognosis.

4. Follow-up. In this step, the evaluation of the effective-ness of the counselling procedure is done and in this, it is observed what the client could achieve through counselling?

5. Prognosis. This step can be a part of diagnosis and can be a separate step too. It depends upon the available specfic information. Where the result of the problem or knowledge of the situation occurs then the prognosis becomes the part of the diagnosis. Both, prognosis and diagnosis are co-operative processes in between the counsellor and the client.

6. Counselling or Treatment. It is that step when the counsellor helps the client. In this, various questions are answered. Client answers these questions for himself, such as, How I can make changes in self? What can be its alternate? If such goes on, how the progress will occur in future?

Advantages of Directive Counselling

1. This method is useful from the time-consumption view-point. It saves a lot of time.
2. In this process, the counsellor becomes readily available to help the client which makes the client very happy.
3. In this type of counselling, there is more focus on the problem and less on the person.
4. Counselling focuses more on the intellectual aspect of the person than his emotional aspect of the personality.

Disadvantages of Directive Counselling

1. In this process the client is more dependent. He is also unable to solve the new problems of mal-adjustment.
2. There is scarcity of informations regarding the pupil which creates the possibility of wrong counselling.
3. As the client is never independent of the counsellor, it is not an effective and best guidance.
4. The counsellor fails in saving the client to commit the mistakes in future.
5. Unless and until a person does not develop some attitudes through experiences he cannot make any decision himself. This type of guidance always lacks such experience and development of attitudes.

Non-Directive or Client Centred or Permissive Counselling

The main advocate of non-directive or client-centred or premissive counselling is Carl R. Rogers. This principle was developed in many years. Hence, various areas got included in such type of counselling such as, the development of personality, group leader-ship, education and learning, creativity, interpersonal relations and the nature of an entirely active person. This theory was developed in between 1930 and 1940. This theory believes that the sufficient means of solving a person's problem are present within the person himself. The function of the counsellor is to provide such an environment in which the client is independent to grow so that he may become what he likes to become. This school of thought gives importance to the emotional aspects of vocational and emotional problems and disapproves the diagnostic information as a part of counselling process.

The client-centred counselling revolves around the client. In this, the client is encouraged to leadin conversation and to express his own attitudes, feelings and ideas. The counsellor mostly remains passive. He never interferes the client's ideas, thoughts, feeling, flow of expressions. The counsellor helps the client in completing his conversation, basically the counsellor makes his efforts to develop rapport and mutual feeling of confidence in both the parties.

In this approach, open-end questions are asked. These questions are loosely structured. In the answers of these questions, the person projects his own personality. The main concern of the counsellor is with the summarization of emotional contents told by the client.

When the client is answering, he should be encouraged to speak in detail through proper methods. The client feels that the counsellor really respects the client's ideas. He gets the impression from the types of questions which a counsellor asks from the client and the interviewer is taking interest in the client. The counsellor does not ask the questions merely to explore the facts. In the non-directive counselling, every person has the right to be independent as a specialist psychologist. In such type of counselling, diagnostic instruments are used either infrequently or completely not used.

Basic Assumptions of Non-Directive Counselling

1. Tendency Towards Actualization. In the earlier writings of Rogers, it was emphasized that the person's or client's growth and developmental capacity is that person's characteristic on which the methods of counselling and psychotherapy depend. Even after so many years, his this opinion got strength that a persons' inherent tendency includes growth, adjustment, socialization; freedom etc. This directional tendency is termed as 'Actualizing Tendency' these days.

2. Belief in the Dignity of Man. Rogers believes in the dignity of the man. He considers the person capable of taking decisions and he accepts his right to do so.

3. Man is Trustworthy—Rogers considers the person basically good and reliable. He also knows that the person behaves in an unreliable manner too. A person is born with some urges which must be controlled if healthy personality development is desired.

4. Man is Wiser than His Intellect. When some organization is functioning independently and effectively, then the 'awareness' is a very small part or component of the entire process. When the organization feels some difficulty in its system of functioning, this awareness becomes more sharpened and focussed.

Snyder has also mentioned the following assumptions of non-directive counselling:

(*i*) The client has right to select the goals of his life.

(*ii*) If the client is given an opportunity, he will select those goals which may possibly provide him great happiness.

(*iii*) In the counselling situation, one should reach very quickly at a point from where the client may be able to start functioning independently.

Steps in Non-Directive Counselling

Carl Rogers has given the following steps of this non-directive counselling:

1. Defining the Problematic Situation. First of all the counsellor should define the problematic situation.

2. Termination of Counselling Situation. The counsellor looks for a point where he can terminate the counselling situation after all the above steps. According to this school of thought, either the client or the counsellor can suggest for such termination of counselling situation.

3. Development of Insight. The counsellor goes on thinking regarding the client's new feelings along with the development of client's insight and he goes on classifying all those new feelings.

4. Classification of Positive and Negative Feelings. After the free expression of feelings by the client, the counsellor identifies his negative and positive feelings and he classifies them.

5. Free Expression of Feelings. After the first step, the client is made aware of the fact that he can express his feelings freely and the counsellor approves this.

Characteristics of Non-Directive Counselling

1. It is a client-centred counselling.
2. In this school of thought, the counsellor is passive to the maximum.
3. In this type of counselling, there is closeness betwen the picture of the self-drawn by the client and a desirable or an ideal picture of his own.
4. In this type of counselling, defensiveness falls down.
5. Through this counselling his psychological adjustment improves.
6. A person is accepted in his originality and he is free to express his any attitude.
7. It is based on the principle that a person has so much capacity and drive that he can grow and develop so that he may face the situations in reality.

8. Its use reduces psychological tensions.
9. Client's behaviour is considered more matured emotionally.
10. This school of thought is just reverse to that of directive counselling.
11. In this counselling, the entire responsibility is of client or the person.
12. In his research Axlines has shown that the reading improvement speeds up as a result of client-centred play way-methods in the primary schools even if special reading instructions are not given.

Disadvantages of Non-Directive Counselling

1. This counselling is not so deep like psycho-analysis.
2. This counselling has a drawback of lack of permission for flexibility to the counsellor.
3. Many counselling situations do not arise successfully in this client-centred counselling.
4. The client cannot depend on his resources, decisions and intellect.
5. It is a time-consuming process. Once the client strats his conversation, he does not cease to speak. This makes the other's turn for seeking counselling impossible.
6. This client-centred theory suffers the basic defect that it does not care how the behaviour is affected by the stimulus-situation and the nature of the environment.
7. It is not possible in all the schools because the counsellor has to attend many pupils.
8. All the problems cannot be solved orally or verbally or by talks only.
9. Sometimes, due to the counsellor's passiveness, the client hasitates in expressing his feelings.

10. The client is allowed for the free expression of his present attitudes but it is not attempted to explain why these attitudes occur? There is no research regarding the past, no suggestion, no attempt of re-education.

Advantages of Non-Directive Counselling

1. The development of the problem-solving ability in the client through this school of thought is definite.
2. Being client-centred school of thought, other unwanted activities and tasts can be avoided.
3. This school of thought shifts the person from his unconscious state of mind to the conscious state of mind and releases tension.
4. This type of counselling leaves its impressions for a longer period.

Organizing School Guidance Service

In order to organize school guidance services, certain principles should essentially be followed. Jones has insisted not to separate guidance from the normal school life. According to him, this can be focussed in some specific part of the subject. It cannot be restricted to the office of the counsellor or a headmaster. Jones has assigned the responsibility of providing guidance to every teacher. In this exercise, co-operation of all is desirable. Hence, for organizing school guidance service, application of certain principles is must.

Humphery and Traxler, in their book entitled 'Guidance Services,' have mentioned some basic things for organizing school guidance service, which are as follows:

Defining the Objectives Clearly. The objectives of entire guidance programme should be clearly defined at the time of organizing school guidance service. While defining these objectives, the school ideals and needs of the pupils must be kept in mind.

Defining the Tasks. All the activities to be performed through the guidance programme must be enlisted beforehand.

Defining the Relations. The relations of employees working in the guidance programme, whether full-time or part-time, must be defined clearly. Inspite of this, their relations should be defined in accordance with their guidance responsibilities with other employees.

Defining the Powers. The workers should be made aware of their powers just like their duties.

Allocation of Duties. The success of the guidance programme depends upon the allocation of duties to the persons involved in this job in advance so that every person should have a clear idea what he is to perform. The allocation of these duties should be according to the abilities of the persons or workers.

Simplicity. The organization of school guidance service should not be of complicated nature. The frame-work of its structure should remain simple as far as it is possible, because every person will start taking interest in such a simple framework of the organization.

Nature of Guidance Organization

It is proper to decide about the structure of school guidance service prior to its beginning such as number of its employees, finance, size etc. The basis of its composition should be the objectives of the institute, financial resources and the number of students in the school etc.

Crow and Crow, in his book, has also suggested the following points to keep in mind before executing the plan of guidance programme:

1. First of all it should be decided the number of persons to be involved in executing the guidance programme and the time required for the same.

2. How much increase in the workers' strength is required.
3. Is the building and the space are enough for conducting the various activities of guidance?
4. Are the teachers and other employees available in the school capable of devoting expected time and energy in the guidance programme?
5. Which teachers are available to provide various proposed services?
6. To what extent other community agencies can be associated with the guidance programme?
7. Do the employees exhibit interest in the programme? If yes, to what extent?
8. What is the viewpoint of schools and the society regarding the expansion of guidance services?
9. In which experience areas the pupils are required to serve?
10. Would the provision of finance in the school budget be possible?
11. How the pupils can be motivated to understand for themselves the value of the guidance programme?

Forms of Organization of Guidance Services

The following can be the forms of organization of guidance services:

1. The Centralised Form. The guidance work should be performed by the experts because the teachers are not trained in this guidance job.

The teachers cannot provide properly any assistance to the pupils who need the same to solve their problems. the central form of the guidance programme means conducting and controlling most of the guidance activities from the central guidance office. All the teachers also perform their duties according to the supervision of Central Guidance Bureau and its orders.

2. Decentralized Form. The decentralized form of the guidance means providing guidance as a responsibility of the teachers. The teacher remains in close contact with the pupils of his class. He understands the needs and problems properly of his pupils. Due to this reason the teacher can help the pupils in a better way. Some people also believe that if a separate department is opened for guidance in the school, the teacher will not show any interest in guidance work and also he will not consider guidance as his responsibility. Hence, it would be better if the responsibility of guidance work in schools lies with the teachers.

3. Mixed Form. The centralised and decentralised form of the guidance have their own merits and demerits. No form of guidance is practically possible independently. Hence, some experts express their views that the form of guidance programme should be mixed one. The mixed form means—guidance provided by the teachers and experts collectively. Crow and Crow also said, supporting the mixed form, that the school guidance programme includes the co-ordinated services of administrators, teachers, employees and social institutions etc. There are certain jobs in guidance programme which the teachers can do better, such as, collection of informations related to the pupils. There are certain jobs which can be done better with the help of experts.

The designs or forms of school guidance service organization can also be described in the following terminology:

(a) The Line Organization.

(b) The Staff Organization.

(c) Mixed Line and Staff Organization.

In the line organization, the employees are arranged in an order with regard to their duties or rights. The top position occupies by the administrative officer *i.e.*, head of the administration. There other assistant officers and the officers and employees working under them come in the sequence. Being head, the administrative officers supervises and examines the performance of other officers and employees.

The administrative officer controls both the line and staff organization. The staff members deputed in certain areas have their definite responsibilities. The specialist employees play a special role in the process of guidance management and policy making even they work under the administrative officer.

The mixed form of line and staff organization is known as 'line-staff mixed organization'. Such organizations have the elements of both types of organizations. The provision of such form of the organization is possible only in big institutions and occurs there.

Persons Working in School Guidance Services and their Functions

As it has already been made clear that the guidance job is a specific task. To run this programme in schools, a specific category of staff members and officers is required. In the school guidance service organization, the participation of the following persons is mainly involved:

1. Headmaster.
2. School Doctor.
3. Counsellor.
4. Class Teachers.
5. School Psychologist.
6. Placement Officer.
7. School Social Worker.
8. School Activities Co-ordinator.
9. School Psychoanalyst.
10. Child Welfare Worker.

Characteristics of Well-Organised Guidance Programmes

The following should be the characteristics of a well-organized guidance programme:

1. Child Centredness. The main characteristic of guidance programme should be its child centredness *i.e.*, the organization of guidance services should be in accordance with the needs, interests of the children and objectives.

2. Specialisation. The role of specialists is very important in the organization of guidance services. Hence, for the success of guidance service, the presence of specialists in the guidance service is essential, such as the counsellor, psychologist, psycho-analyst, social worker of the school, doctor etc.

3. Budgetary Provisions. Budgetary provisions are also necessary to make the guidance service better. Budgetary provisions are made before starting the programme.

4. Continuity. Continuity of each guidance service is very essential. Continuity of problems occurs in the life of each person. Therefore, the continuity of these services is needed.

5. Proper Information and Record. The superiority of the services also depends upon the proper information and record. The guidance service should be used to collect the informations about educational and vocational needs and tests and to record them.

6. Co-operation. The characteristic of a good guidance service organization is that it should seek the co-operation of the parents. The guidance service can be made better by seeking the interest and co-operation of each staff member.

7. Co-ordination. There should be another characteristic of guidance service that it should maintain co-ordination between all the associate agencies and the community for the guidance of youth for a longer period.

8. Guidance for All. The principle of guidance for all is followed in a good guidance service. In a democratic country each pupil has a right to get guidance. In school, all the pupils should be provided with guidance.

9. Selection of Guidance Workers. For a good guidance programme, the guidance workers should be selected on the basis of their abilities, training and experiences. It is not possible to perform all the duties related to guidance by the teacher alone. Hence, for specific duties in guidance, persons with special abilities in guidance are needed to be appointed.

10. Confidential. It is very important to maintain confidentiality of the collected informations for the guidance in guidance service so that the confidence of the pupils on the guidance workers may be maintained. After winning the confidence of the pupils, true informations about them can be collected. Only the counsellor should have access to the confidential informations.

11. Opportunity for Final Decision to the Client. In a good guidancc programme, the opportunity for final decision should be provided to the client. There should be no compulsion on them.

12. Sufficient Time for Guidance. The guidance workers should be allotted sufficient time to provide successful guidance in the guidance service, such as time for testing and time for counselling. In time-table too, the sufficient time should be allotted for guidancc.

13. Use of Various Methods. It is not possible to apply a single method for collecting informations meant for guidance. Also, it is not proper to use single method for the same purpose. Only one method cannot collect reliable informations. It is essential to apply various methods on one person at the different occasions and in different situations. It increases the reliability of the programme.

14. Simplicity. If the guidance programme is simple, it will be interesting and successful. In such simple programmes, all the persons show more interest.

15. Finding Solution. The utility of guidance service increases when it is organised for finding the causes of the problems. These problems should not be allowed to arise and preparation should be made to solve them if the same arise.

16. Responsibilities According to Abilities. A good guidance organization is that where all the persons are assigned their responsibilities according to their abilities and capacities. If they are not assigned the responsibilities according to their abilities, then the success of the entire guidance programme cannot be expected. However, the client can experience some harm.

17. Providing Opportunities for Acquiring Abilities. The opportunities should be provided to the guidance of workers and other persons associated with such programme to increase their abilities. It is essential because new methods are innovated in the field of guidance every year and old methods are modified and improved. Every year new tests are constructed. The guidance worker should be given an opportunity to remain in-touch with all these modifications or changes.

Hence, we see that the above characteristics are responsible for the success of guidance service in an organization.

Index

J

K

L

M

N

O

P